ENVIRONMENTAL REMEDIATION AND MANAGEMENT OF TRENCHES CONTAINING HISTORIC RADIOACTIVE WASTES

The following States are Members of the International Atomic Energy Agency:

AFGHANISTAN
ALBANIA
ALGERIA
ANGOLA
ANTIGUA AND BARBUDA
ARGENTINA
ARMENIA
AUSTRALIA
AUSTRIA
AZERBAIJAN
BAHAMAS
BAHRAIN
BANGLADESH
BARBADOS
BELARUS
BELGIUM
BELIZE
BENIN
BOLIVIA, PLURINATIONAL STATE OF
BOSNIA AND HERZEGOVINA
BOTSWANA
BRAZIL
BRUNEI DARUSSALAM
BULGARIA
BURKINA FASO
BURUNDI
CABO VERDE
CAMBODIA
CAMEROON
CANADA
CENTRAL AFRICAN REPUBLIC
CHAD
CHILE
CHINA
COLOMBIA
COMOROS
CONGO
COOK ISLANDS
COSTA RICA
CÔTE D'IVOIRE
CROATIA
CUBA
CYPRUS
CZECH REPUBLIC
DEMOCRATIC REPUBLIC OF THE CONGO
DENMARK
DJIBOUTI
DOMINICA
DOMINICAN REPUBLIC
ECUADOR
EGYPT
EL SALVADOR
ERITREA
ESTONIA
ESWATINI
ETHIOPIA
FIJI
FINLAND
FRANCE
GABON
GAMBIA, THE
GEORGIA
GERMANY
GHANA
GREECE
GRENADA
GUATEMALA
GUINEA
GUYANA
HAITI
HOLY SEE
HONDURAS
HUNGARY
ICELAND
INDIA
INDONESIA
IRAN, ISLAMIC REPUBLIC OF
IRAQ
IRELAND
ISRAEL
ITALY
JAMAICA
JAPAN
JORDAN
KAZAKHSTAN
KENYA
KOREA, REPUBLIC OF
KUWAIT
KYRGYZSTAN
LAO PEOPLE'S DEMOCRATIC REPUBLIC
LATVIA
LEBANON
LESOTHO
LIBERIA
LIBYA
LIECHTENSTEIN
LITHUANIA
LUXEMBOURG
MADAGASCAR
MALAWI
MALAYSIA
MALI
MALTA
MARSHALL ISLANDS
MAURITANIA
MAURITIUS
MEXICO
MONACO
MONGOLIA
MONTENEGRO
MOROCCO
MOZAMBIQUE
MYANMAR
NAMIBIA
NEPAL
NETHERLANDS, KINGDOM OF THE
NEW ZEALAND
NICARAGUA
NIGER
NIGERIA
NORTH MACEDONIA
NORWAY
OMAN
PAKISTAN
PALAU
PANAMA
PAPUA NEW GUINEA
PARAGUAY
PERU
PHILIPPINES
POLAND
PORTUGAL
QATAR
REPUBLIC OF MOLDOVA
ROMANIA
RUSSIAN FEDERATION
RWANDA
SAINT KITTS AND NEVIS
SAINT LUCIA
SAINT VINCENT AND THE GRENADINES
SAMOA
SAN MARINO
SAUDI ARABIA
SENEGAL
SERBIA
SEYCHELLES
SIERRA LEONE
SINGAPORE
SLOVAKIA
SLOVENIA
SOMALIA
SOUTH AFRICA
SPAIN
SRI LANKA
SUDAN
SWEDEN
SWITZERLAND
SYRIAN ARAB REPUBLIC
TAJIKISTAN
THAILAND
TOGO
TONGA
TRINIDAD AND TOBAGO
TUNISIA
TÜRKİYE
TURKMENISTAN
UGANDA
UKRAINE
UNITED ARAB EMIRATES
UNITED KINGDOM OF GREAT BRITAIN AND NORTHERN IRELAND
UNITED REPUBLIC OF TANZANIA
UNITED STATES OF AMERICA
URUGUAY
UZBEKISTAN
VANUATU
VENEZUELA, BOLIVARIAN REPUBLIC OF
VIET NAM
YEMEN
ZAMBIA
ZIMBABWE

The Agency's Statute was approved on 23 October 1956 by the Conference on the Statute of the IAEA held at United Nations Headquarters, New York; it entered into force on 29 July 1957. The Headquarters of the Agency are situated in Vienna. Its principal objective is "to accelerate and enlarge the contribution of atomic energy to peace, health and prosperity throughout the world".

IAEA NUCLEAR ENERGY SERIES No. NW-G-3.3

ENVIRONMENTAL REMEDIATION AND MANAGEMENT OF TRENCHES CONTAINING HISTORIC RADIOACTIVE WASTES

LEGACY TRENCH SITES

INTERNATIONAL ATOMIC ENERGY AGENCY
VIENNA, 2025

Publishing Section
International Atomic Energy Agency
Vienna International Centre
PO Box 100
1400 Vienna, Austria
tel.: +43 1 2600 22529 or 22530
email: sales.publications@iaea.org
www.iaea.org/publications

Printed by the IAEA in Austria
February 2025
STI/PUB/2075
https://doi.org/10.61092/iaea.rgwy-0qia

IAEA Library Cataloguing in Publication Data

Names: International Atomic Energy Agency.
Title: Environmental remediation and management of trenches containing historic radioactive wastes : Legacy trench sites. / International Atomic Energy Agency.
Description: Vienna : International Atomic Energy Agency, 2025. | Series: IAEA nuclear energy series, ISSN 1995-7807 ; no. NW-G-3.3 | Includes bibliographical references.
Identifiers: IAEAL 24-01730 | ISBN 978-92-0-153823-9 (paperback : alk. paper) | ISBN 978-92-0-154023-2 (pdf) | ISBN 978-92-0-153923-6 (epub)
Subjects: LCSH: Radioactive wastes. | Radioactive wastes — Environmental aspects. | Radioactive wastes — Management. | Radioactive waste sites. | Hazardous waste site remediation. | Radioactive waste disposal.
Classification: UDC 621.039.7 | STI/PUB/2075

FOREWORD

The IAEA's statutory role is to "seek to accelerate and enlarge the contribution of atomic energy to peace, health and prosperity throughout the world". Among other functions, the IAEA is authorized to "foster the exchange of scientific and technical information on peaceful uses of atomic energy". One way this is achieved is through a range of technical publications including the IAEA Nuclear Energy Series.

The IAEA Nuclear Energy Series comprises publications designed to further the use of nuclear technologies in support of sustainable development, to advance nuclear science and technology, catalyse innovation and build capacity to support the existing and expanded use of nuclear power and nuclear science applications. The publications include information covering all policy, technological and management aspects of the definition and implementation of activities involving the peaceful use of nuclear technology. While the guidance provided in IAEA Nuclear Energy Series publications does not constitute Member States' consensus, it has undergone internal peer review and been made available to Member States for comment prior to publication.

The IAEA safety standards establish fundamental principles, requirements and recommendations to ensure nuclear safety and serve as a global reference for protecting people and the environment from harmful effects of ionizing radiation.

When IAEA Nuclear Energy Series publications address safety, it is ensured that the IAEA safety standards are referred to as the current boundary conditions for the application of nuclear technology.

Early radioactive waste disposal practices typically involved trenches excavated directly into the soil or with simple concrete liners. Little consideration was given to the environmental setting or to conditioning the waste before disposal. Such trench sites exist in many Member States and, owing to either their original design or failed integrity over time, many of these sites may now pose a risk to people and the environment.

The Network of Environmental Management and Remediation (ENVIRONET) is an internationally recognized and active community facilitated by the IAEA. The aim of ENVIRONET is to support Member States in addressing risks posed to people and the environment from contaminated sites. In 2014, ENVIRONET participants identified a specific need to support Member States with the management and remediation of trench sites resulting from past activities. These sites are often termed 'legacy trench sites'; hence the project was named the LeTrench Project.

The IAEA convened four Technical Meetings, including site visits, to enable information and knowledge transfer, peer support and advice, and collection of information to support this publication. The IAEA expresses its appreciation to the Member States and organizations that have supported Technical Meetings with site visits.

This publication draws together the experience of Member States in the evaluation, management and remediation of legacy trench sites. It discusses the activities necessary to address the potential risks posed by legacy trench sites and provides Member State examples related to the implementation of these activities. This publication concludes with Member State case studies that provide an overview of the history, challenges and activities at various legacy trench sites.

The IAEA would like to thank the experts who contributed to the drafting and review of this publication.

CONTENTS

1. INTRODUCTION

1.1. BACKGROUND

The research and application of nuclear technologies started in the 1940s with work on atomic weapons and the development of research reactors and then the first generation nuclear power plants began to operate. In those early days, the approach to managing radioactive waste was often simple, and in many cases used trenches. The construction of these trenches varied; some were engineered but others were simple excavations into the soil. In some instances, waste was packaged, but in others, it was tipped directly into the trench. Trenches were also used in the aftermath of accidents, where there was little time to engineer alternative waste management options.

In general, these historic activities were not subject to regulatory control or, if regulated, were not in accordance with today's national or international standards, including those formulated by the IAEA. Owing to various shortcomings with their siting, construction, operation (including unconditioned wastes and disused sealed radioactive sources) or other factors, such trenches may pose a risk to people and the environment. These sites are collectively referred to as 'legacy trench sites'.

1.2. OBJECTIVE

The objective of this publication is to describe the overall process necessary to facilitate the environmental management and remediation of a legacy trench site. The publication elaborates on the characterization, assessment and potential remediation of such sites and provides guidance on management aspects such as decision making and engaging interested parties. It is important that these tasks are implemented having regard to the environmental, radiological, regulatory and societal context of the site. A key aim is to highlight specific issues that make legacy trench sites different from other contaminated sites and to draw the attention of relevant stakeholders to the issues surrounding such sites.

The publication also provides case studies from various countries to illustrate legacy trench sites that have been or are in various stages of assessment and remediation. Relevant experience and lessons learned for these sites are provided throughout this publication.

Guidance provided here, describing good practices, represents expert opinion but are not made on the basis of a consensus of Member States.

1.3. SCOPE

The publication addresses sites where there is concern that past activities to dispose of radioactive waste (raw and/or conditioned) into near surface excavations in the ground (with or without engineered containment) may pose a risk to people and/or the environment. Hence, these sites require assessment, ongoing management and possibly remediation.

This publication does not evaluate the adequacy of existing or past waste disposal practices. It is not intended for assessing or managing currently operating disposal sites, although it may be applicable to non-operational parts of such sites.

If remediation is necessary, this may involve waste retrieval, then predisposal and disposal activities of the retrieved wastes. This publication does not provide guidance on the waste management aspects of a legacy trench site, but does provide references to the appropriate international guidance and some relevant case studies.

The typical characteristics of a legacy trench site are outlined in Section 1.1, but all sites are different. It is advised that the reader uses discretion to determine whether a specific site fits within the scope of this publication. Even if a site is not considered a legacy trench site, the content of this publication may nevertheless be useful.

1.4. STRUCTURE

Section 2 establishes the importance of the topic of legacy trench sites, identifies the unique characteristics of these sites and provides the overall process for managing legacy trench sites. Another issue discussed in this section is that there is often limited specific regulatory guidance for legacy waste trenches. This may be partly due to the limited number of these sites per country and whether the site is considered as existing exposure situation or a planned exposure situation.

Section 3 provides information on the potential waste inventory of these sites and discusses trench location, construction and waste emplacement.

Section 4 discusses activities that may be undertaken to evaluate the site, including reconstruction of the site history, site characterization, and safety assessment and modelling. This section provides information on the common approaches and technologies used and highlights where these have been used by Member States.

Using the site evaluation information described in Section 4, Section 5 discusses decision making to support ongoing management and identification of the site end state. Defining the site end state is much broader than just identifying a remediation target for soil and/or groundwater.

Section 6 highlights the different remediation approaches and technologies that could be applied to legacy trench sites. It is important to note that improved management of a legacy trench site, rather than waste retrieval, may be sufficient.

Good practice recognizes the importance of engaging interested parties in the management or remediation of legacy trench sites. Section 7 discusses the potential concerns that may be raised by interested parties and the benefits and challenges of including interested parties in the decision making process.

Section 8 discusses record management, noting that a major challenge for legacy trench sites is often the lack of data and information due to poor historic knowledge management practices. Thus, it is important that any evaluation or remediation of a site is well documented.

Section 9 presents the key conclusions and lessons learned by Member States from the environmental remediation and management of legacy trench sites.

Appendix I provides seven case studies, which identify the key challenges and lessons learned from the evaluation, management and remediation of legacy trench sites. Appendix II provides further detailed information on key exposure pathways and scenarios for legacy trench sites for risk assessment and modelling.

2. OVERVIEW OF LEGACY TRENCH SITES

Legacy trench sites exist in many countries. Early land based disposal practices typically involved tumble-tipping of unpacked waste into shallow unlined trenches. This practice was undertaken at the Low Level Waste Repository (LLWR), United Kingdom (UK), in its early years (Fig. 1), and also at the Chalk River site, Canada. Subsequently, wastes were often packaged in a variety of container types and either randomly placed or stacked into the trenches. Examples of this approach can be found at the Little Forest Legacy Site (LFLS) in Australia, the Ezeiza site in Argentina (Fig. 2) and the Savannah River site in the United States of America (USA). Trenches were often backfilled using materials removed during trench excavation, compacted and graded to create an earthen mound cap to promote water runoff and minimize water infiltration and ponding.

In general, the waste in legacy trench sites originated from research activities or the operation of research reactors and first generation nuclear power plants. However, trenches have also been used to manage waste resulting from accident situations, for example at Chornobyl site, Ukraine, and at Windscale (Sellafield) site, UK.

FIG. 1. Historic waste being placed at the Low Level Waste Repository, United Kingdom. Photograph courtesy of Low Level Waste Repository Ltd.

FIG. 2. Drummed historic waste, adjacent to covered trenches containing other drummed historic waste, at the Ezeiza site, Argentina. Photograph courtesy of Centro Atómico Ezeiza.

The management of wastes resulting from the accident at Unit 4 of the Chornobyl nuclear power plant (ChNPP) used a mix of trenches and above ground mounds, known as 'clamps' in Ukraine. The mounds were formed from bulldozed heaps of waste covered by soil. Figure 3 shows an area in the Red Forest burial site comprising trenches and clamps. However, owing to seasonal flooding, the trenches are submerged and only the clamps can be seen.

Across the countries of the former Union of Soviet Socialist Republics (USSR), a standardized trench disposal facility was used, referred to as a

FIG. 3. Seasonal flooding at the Red Forest radioactive waste burial site, Chornobyl Exclusion Zone, Ukraine. Photograph courtesy of D. Bugai, Institute of Geological Sciences, Kiev.

'RADON' type facility (Fig. 4)[1]. These facilities are so named because their designs were based on the same concept as that of the two central facilities, near Moscow and St Petersburg, operated by the Scientific and Industrial Association RADON. In some countries, such as Ukraine and the Republic of Moldova, RADON type facilities are now considered to be storage facilities rather than disposal facilities. The design of RADON type facilities comprised prefabricated concrete vaults (of 3–4 m depth) and shallow borehole type storage for disused sealed radioactive sources. Waste material was either loose or packaged in a variety of container types, randomly positioned or stacked into the subsurface vaults. Work is ongoing to assess the RADON type facilities to determine if they are suitable for waste storage, if they can be retained as disposal facilities, or if waste retrieval and/or environmental remediation is necessary.

Table 1 summarizes the key characteristics of legacy trench sites from several countries. Figure 5 depicts the operational periods of the sites listed in Table 1. The peak period of operations was from the 1960s to the 1990s. The

Text cont. on p. 12.

[1] Thirty-six RADON type sites exist: sixteen in the Russian Federation; five in Ukraine; one each in the former USSR countries of Armenia, Azerbaijan, Belarus, Estonia, Georgia, Kazakhstan, Kyrgyzstan, Latvia, Lithuania, Republic of Moldova, Tajikistan, Turkmenistan and Uzbekistan; and one each in Bulgaria and Hungary.

FIG. 4. Special Facilities 5101 and 5102 of GIES (a RADON type facility) at Chisinau, Republic of Moldova. Photograph courtesy of I. Gisca, National Radioactive Waste Management Company.

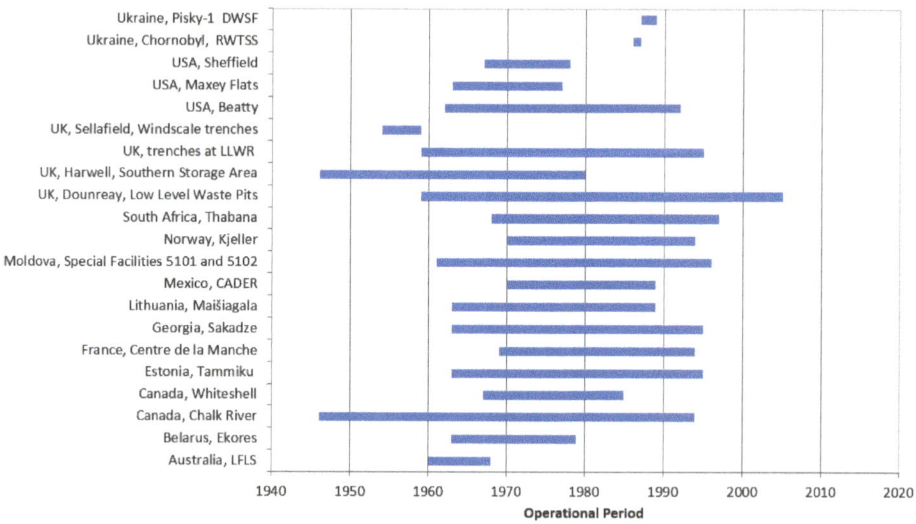

FIG. 5. Operational periods of legacy trench sites. DWSF — decontamination waste storage facility; CADER — Centro de Almacenamiento de Desechos Radiactivos; LFLS — Little Forest Legacy Site; LLWR — Low Level Waste Repository; RWTSS — radioactive waste temporary storage site.

TABLE 1. SUMMARY OF KEY CHARACTERISTICS FROM SEVERAL LEGACY TRENCH SITES

Country, site	Operational phase	Construction	Known environmental impact	Site status
Argentina, Ezeiza	Trench T1: 1969–1988 Trench T2: 1988–1999	T1 trench: unlined trench; multilayer cap. T2 trench: unlined trench; concrete side walls; compacted silt base; sand backfill and multilayer cap.	Localized contaminant migration into groundwater; tritium and uranium plume within the site boundary	Evaluation and planning.
Australia, Little Forest Legacy Site	1960–1968	Unlined trenches; soil cap.	Surface contamination (bath-tubbing[a]); tritium plume	Evaluation and planning.
Belarus, Ekores	1963–1979	RADON type facility.	None identified by monitoring activities	Ongoing monitoring and evaluation. Awaiting end state decision on waste retrieval.
Canada, Chalk River	1946–1994	Unlined trenches and pits; soil cap, concrete lined trenches and bunkers; concrete cap, asphalt lined and capped trenches.[b]	Contaminant migration into soils and groundwater; tritium and strontium-90 plumes	Evaluation, planning and remediation.
Canada, Whiteshell	1967–1985	Unlined trenches with natural clay cap.	Limited migration of contaminants	Evaluation, planning and remediation.

TABLE 1. SUMMARY OF KEY CHARACTERISTICS FROM SEVERAL LEGACY TRENCH SITES (cont.)

Country, site	Operational phase	Construction	Known environmental impact	Site status
Estonia, Tammiku	1963–1995	RADON type facility.	Environmental modelling identified impact to people and the environment	Waste retrieved; structures decontaminated and demolished. Remediation planned for 2022.
France, Centre de la Manche	1969–1994	Unlined trenches; soil cap. Concrete lined trenches; sand or concrete backfill.	Tritium migration to local river	Post-remediation management.
Georgia, Saakadze	1963–1995	RADON type facility.	Safety assessment concluded no impact to people	Post-remediation management.
Lithuania, Maišiagala	1963–1989	RADON type facility. Additional cap installed in 2006 comprises sand, membrane and soil.	Tritium plume	Evaluation and planning.
Mexico, CADER	1970–1989	Unlined trenches with concrete slab cap for radioactive waste. Reinforced concrete/brick walled vaults for disused sealed radioactive sources.	Little identified by characterization activities	Ongoing monitoring, whilst awaiting end state decision.

TABLE 1. SUMMARY OF KEY CHARACTERISTICS FROM SEVERAL LEGACY TRENCH SITES (cont.)

Country, site	Operational phase	Construction	Known environmental impact	Site status
Republic of Moldova, Special Facilities 5101 and 5102[c]	1961–1996	RADON type facility.	Risk assessment identified impact to people	Evaluation and planning.
Norway, Kjeller	1970–1994	Unlined trenches; 1.5–2 m clay and soil cap.	None identified during remedial works	Waste retrieved; site remediated (2002).
South Africa, Thabana (Radiation Hill)	1968–1997	Unlined trenches.	None currently identified	Ongoing monitoring and planning. Remediation strategy with the regulatory body for approval.
Ukraine, Chornobyl Exclusion Zone, radioactive waste temporary storage sites including the Red Forest site	1986–1987	Unlined trenches and above ground mounds; soil cap.	Contaminant migration into soils, groundwater and accumulation in vegetation	Evaluation. Maintenance of exclusion zone controls and selective retrieval of wastes.
Ukraine, Pisky-1 Decontamination Waste Storage Facility	1987–1989	Unlined trench; soil cap.	Safety assessment identified the site posed a potential risk to the public	Remediated (2019).

TABLE 1. SUMMARY OF KEY CHARACTERISTICS FROM SEVERAL LEGACY TRENCH SITES (cont.)

Country, site	Operational phase	Construction	Known environmental impact	Site status
UK, Dounreay, Low Level Waste Pits	1959–2005	Unlined pits lying on bedrock; soil cap.	None currently identified	Evaluation and planning; current assumption is waste will be retrieved.
UK, Harwell, Southern Storage Area	1946–1980	Unlined trenches and pits.	Contaminant migration into soils and groundwater	Remediated (2002–2003); unrestricted use.
UK, trenches at the Low Level Waste Repository	1959–1995	Trenches — unlined trenches in low permeability clay or lined with bentonite; geomembrane and soil interim cap.	Contaminant migration into soils and groundwater; tritium plume	Planning and post-operational management.
UK, Sellafield, Windscale Trenches	1954–1959	Unlined trenches; originally a soil cap but capped with tarmac (c. 1991) and repaired between 2014 and 2018.	Tritium plume	Evaluation, planning and interim post-remediation management.
USA, Beatty	1962–1992	Unlined trenches; clay and soil cap. Upgraded soil and gravel cap installed in 1992.	Tritium plume	Evaluation and planning.
USA, Maxey Flats	1963–1977	Unlined trenches; soil cap. Leachate management and new cap comprising soil and geomembranes installed in 2016.	Contaminant migration into soils and groundwater; surface contamination (bath-tubbing)	Remediated 2016 and site closed. Post-remediation management.

TABLE 1. SUMMARY OF KEY CHARACTERISTICS FROM SEVERAL LEGACY TRENCH SITES (cont.)

Country, site	Operational phase	Construction	Known environmental impact	Site status
USA, Sheffield	1967–1978	Unlined trenches. Low permeability clay cap with soil cover installed in 1979.	Contaminant migration into soils, groundwater and off-site lake; tritium and carbon-14 plumes	Post-remediation management.

[a] The collection and overflow of groundwater within the trenches.

[b] Tile holes were used for high level wastes. Tiles holes are subsurface vertical concrete pipes set on a poured concrete base. Some tile holes included a steel liner.

[c] Special Facilities 5101 and 5102 are located at the Central Radioactive Waste Disposal Facility, Chisinau.

short operational phases for the sites containing waste resulting from accident situations (i.e. Chornobyl and Windscale (Sellafield)) are notable.

Appendix I provides case studies for the following legacy trench sites:

— Semi-containment System for Solid Radioactive Waste, Ezeiza Radioactive Waste Management Area (Área de Gestión Ezeiza; AGE), Ezeiza Atomic Centre (Centro Atómico Ezeiza; CAE), Argentina;
— Little Forest Legacy Site, Australia;
— Chalk River Laboratories, Canada;
— Centro de Almacenamiento de Desechos Radiactivos, Mexico;
— Historic Waste Trenches, LLWR, UK;
— Red Forest Radioactive Waste Temporary Storage Site, Ukraine;
— Maxey Flats Disposal Site, USA.

2.1. DRIVERS FOR ADDRESSING REMEDIAL ACTIONS TO LEGACY TRENCH SITES

The key issue and reason to remediate a legacy trench site is the potential risk posed by the migration of contaminants from the waste to people and the environment. There may also be risks associated with accidental or deliberate intrusion into the waste trenches. However, other drivers, of either a technical or non-technical nature, may result in corrective actions. Figure 6 shows the drivers that are likely to be relevant in initiating or determining actions at a legacy trench site. As would be expected, these drivers correspond to those driving the implementation of remediation at contaminated land sites and can form constraints when defining the site end state.

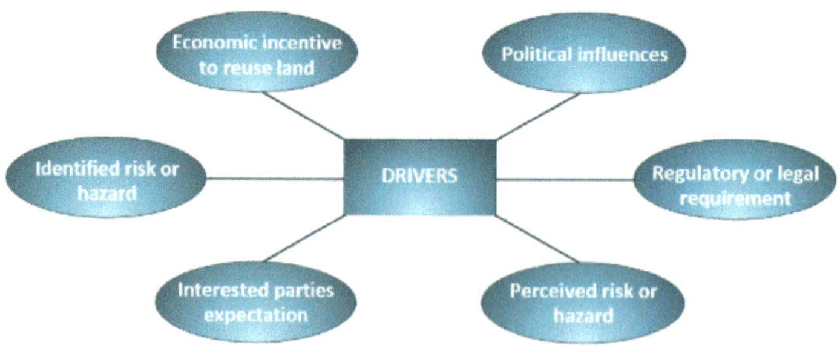

FIG. 6. Drivers for corrective actions to legacy trench sites.

2.2. COMMON CHARACTERISTICS AND ISSUES

Although all legacy trench sites are different, they have many common characteristics and issues, primarily associated with liability, construction, operation, waste types and the associated contaminant inventory and environmental impact. There are uncertainties associated with each of these characteristics, which makes assessing, remediating and defining the end state for these sites difficult. The common characteristics of legacy trench sites are listed below.

Liability

- Assessment is needed to establish whether the current configuration of the waste is suitable for long term management to protect human health and the environment, both now and in the future.
- Typically, responsibility for and ownership of the site are assumed to either belong to the incumbent operator or have been transferred to the government. In addition, the site operator and/or liability owner may have changed since the trenches were initially constructed, or during the operational period or after operations have ceased.
- The approach to regulation or the regulatory status of the site and/or the trenches might not be well defined. This is an issue that still affects legacy trench sites today.
- Understanding the liability can be difficult owing to shortcomings in information and knowledge management. Often the lifetime of the site outlives organizations, personnel and data management systems.
- In some cases, there can be an absence of physical restrictions to control access to the site.

Construction and operation

- The site location was generally selected for convenience rather than the location being a suitable hydrogeological and environmental setting.
- The trench area or site may be part of or adjoin an operational nuclear site or radioactive waste site.
- The trenches are typically near surface excavations (e.g. ditch, channel, pit, vault) in the ground, which might or might not include engineered containment.
- Typically, waste was roughly placed or tipped into the trench without waste processing.

— Subsidence can be a problem for trenches where waste was tumble-tipped or contained a high volume of degradable waste. This can impact the stability of capping layers after the trenches have been closed.
— Often, operational practices were not documented and may have changed over the operational period. For example, the practice of careful emplacement of packaged waste may have changed to the tumble-tipping of unsorted wastes as more wastes were generated.
— Waste acceptance criteria may have not originally existed and/or might have changed over time. Hence, wastes that were acceptable at the beginning of the operational period might not have been acceptable at a later date.
— Operational changes may have occurred as a result of modification of the regulatory approach or the regulatory status of the site and/or trenches.
— Often, record keeping was inadequate during the initial operation of the trenches. Disposal records are often poor or incomplete, and the precise disposal locations of individual waste objects are generally unknown. Historic records were largely handwritten on paper, which has subsequently deteriorated or been lost.
— Environmental monitoring of the site may not have occurred or was carried out only to a limited extent. Monitoring records may also be incomplete or missing.

Types of waste

— The trenches may contain waste associated with the following:
  - Early nuclear and atomic weapon research;
  - Operation of early research reactors and first generation nuclear reactors;
  - Development of reprocessing facilities;
  - Contamination of soils resulting from leaks or spills;
  - Disused sealed radioactive sources;
  - Accident situations.
— The trenches commonly contain a variety of materials (e.g. soils, metals, plastics, rubble, glass), which results in a highly variable and heterogeneous waste.
— Wastes were often poorly characterized.
— Wastes were often not processed and could be liquids or solids.
— Radioactive wastes could be mixed with chemically hazardous and biologically hazardous wastes.

- — The construction of trenches, and/or their failure and/or their lack of waste processing may have allowed the migration of contaminants from the waste to the environment. Hence, legacy trench sites may represent an ongoing source of contamination in the environment.
- — Tritium plumes are common at legacy trench sites.
- — Where controls are not in place, disturbance of the trench and/or waste can occur via human intrusion and/or biointrusion (e.g. plant roots, burrowing animals).
- — The generation of ground gas can occur through the degradation of wastes (e.g. methane from plant material).
- — Surface contamination can occur through the process of 'bath-tubbing', where water accumulated in the trenches overflows.
- — Contamination of the surrounding environment may have occurred during the waste emplacement activities (e.g. surface cross-contamination, liquid spillages).
- — When performing the safety assessment, it can be difficult to define the source term (inventory). Hence, contaminant transport modelling can be difficult, leading to greater uncertainties and conservatism in risk assessment and contaminated soil waste estimates.

2.3. LOCATION AND INVENTORY UNCERTAINTY

Legacy trench sites can be grouped on the basis of the level of information that is known regarding their location and inventory. This grouping can provide initial guidance on the approach and level of effort needed to support further decision making about the site. The groups are as follows:

(1) A trench whose location is known, but where there is only partial or no inventory data.
(2) A trench whose existence is inferred from either existing records or anecdotal evidence. In this instance, there will be uncertainty around both the location and likely inventory.
(3) A trench that has been completely forgotten and may be discovered only by chance (either through records or site characterization projects).

For legacy trench sites in group 1, additional data collection will focus on understanding the types of waste and the potential contaminant inventory. Where waste characterization data are unavailable, the inventory may be inferred from information on the processes or activities occurring at the site when the waste was generated. If the data are uncertain or limited, then site characterization will be needed to support future decisions on the management of the trench. A good example of this category of legacy trench site is the Windscale Trenches located at the Sellafield site, UK. The approximate location of the trenches has been known for over 70 years, as well as some information of the inventory. However, a recent non-intrusive site investigation was necessary to confirm the exact configuration of the trenches.

For legacy trench sites in group 2, the existence of the trench is known, but data are needed to confirm the location and the inventory. Their existence is likely to be inferred from anecdotal information, rather than records. These trenches are likely to be unauthorized or may be present on sites where there was little or no regulatory oversight during operation. Occasionally, these types of trench site are used in an accident scenario, where there is little time to document the location or the inventory of the trench. For these sites, it is likely that both non-intrusive and intrusive site investigation will be needed, as it is unlikely that sufficient existing data will be obtained from other sources.

Legacy trench sites in group 3 do exist, and they are more likely to be present on sites where there was no regulatory oversight during operations. In these cases, corporate memory of the trench has been lost, and therefore no work is being undertaken to find the trench. This type of trench is likely to remain forgotten until it is discovered accidentally or new information becomes available. A recent example of an unknown trench that was discovered is the Veselovsky Pit legacy trench site, located in the Kirovohrad region in Ukraine. In 2017, trench burials were discovered by scrap metal scavengers who excavated steel objects, waste and soil, which were contaminated with caesium-137. The contaminated material is believed to have originated from an accident with disused sealed radioactive sources at an enterprise in Kropyvnytskyi City (former Kirovohrad) in 1988. However, institutional memory and control over the burial site were lost until the contaminated site was rediscovered [1].

Legacy trench sites in groups 2 and 3 present the greatest challenge because it is difficult to establish a plan and budget for their management owing to the many uncertainties and unknown factors.

2.4. PROCESS FOR MANAGING LEGACY TRENCH SITES

Figure 7 sets out the series of key activities that occur at a contaminated land[2] site to address potential risks to people and the environment. In accordance

[2] 'Land' is used as an inclusive term to include ground, groundwater and surface water.

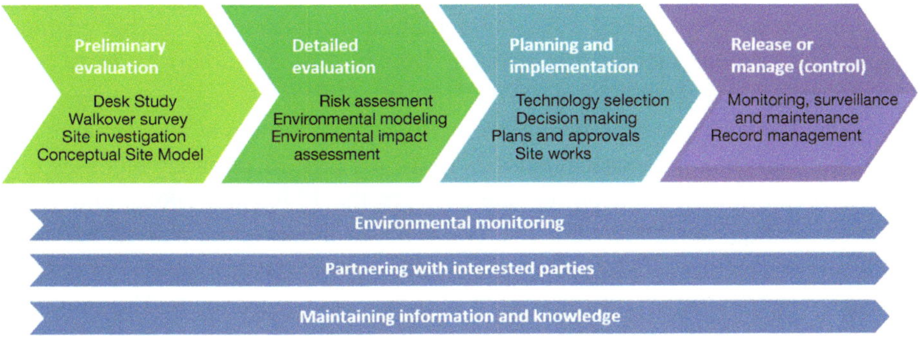

FIG. 7. Phases and key activities in the remediation process for contaminated sites.

with the IAEA Safety Standards Series No. GSG-15, Remediation Strategy and Process for Areas Affected by Past Activities and Events [2], the activities can be broadly grouped into four phases of work: site evaluation, planning, implementation and post-implementation management.

Each phase of work involves several activities, which may relate to only one phase or occur over several phases. Some activities will be iterative and adaptive, depending on the information gathered during previous activities; for example, the creation and update of the conceptual site model.

The phases and activities are generally the same for legacy trench sites, as they are for contaminated land sites. However, because of the characteristics set out in Section 2.2, there are some differences. The remainder of this publication discusses these differences and provides specific guidance on activities where additional considerations are necessary for legacy trench sites.

Recommendations on the remediation process can be found in GSG-15 [2]. Guidance on the strategic and technical aspects of planning and implementing waste management activities can be found in Ref. [3]. Information regarding the selection of the site end state can be found in Ref. [4].

As can be seen in Table 1, there are legacy trench sites at various stages of the process. Many sites are in the site evaluation and planning phases of the remediation process, and most sites have undergone characterization of the environment that surrounds the trenches. For example, at the LFLS, Australia, site characterization began in 1959, prior to the disposal operations. The site characterization work completed in the last 15 years has provided the scientific basis for decisions on the management approach for the site.

At some sites, environmental modelling and safety assessments have demonstrated that ongoing management, perhaps with some physical works, is an appropriate action (e.g. LLWR, UK; State Enterprise Radioactive Waste

Site, Bulgaria). For other sites, remediation activities are being planned (e.g. RADON type facility at Chisinau, Republic of Moldova) or have been completed (e.g. Southern Storage Area, Harwell, UK).

The Harwell site was originally a Royal Air Force site and became the UK's Atomic Energy Establishment in 1946. A range of research facilities and nuclear reactors were built on the site between the 1940s and 1960s, and many of these facilities operated until the 1990s. Decommissioning and remediation of the site started in the late 1990s and is still ongoing. Over a quarter of the site has been released for use as a Science and Technology Campus. The Southern Storage Area was located to the south of the main site and was initially used as an ammunitions store. From 1946, the Southern Storage Area was used for a variety of waste handling, storage and disposal activities. Waste disposal occurred in unlined pits constructed into the chalk geology. Waste comprised hazardous and radioactive contaminated materials originating from laboratory research; laboratory equipment; drummed liquid wastes, including chlorinated hydrocarbons; building rubble; and soils. Following the assessment of potential risks posed by the site and opportunities to reuse the site, the site was remediated between 1994 and 2002. Figure 8 shows the site during the remediation phase; double skinned tents were used to ensure protection of the local population and stop the spread of further contamination. A pump and treat plant was used to remediate the groundwater in the underlying chalk aquifer, which had been impacted by the chlorinated hydrocarbons disposed of at the site. This remediation was a success, although a number of challenges were encountered during the process. For example, the overall non-radioactive waste volumes were 50% higher than those estimated from the characterization information. In addition, the amount of munitions encountered was much higher than expected, which meant that the remediation took longer and was more costly. The site was released from regulatory control and has now been reused for housing.

In the USA, the Maxey Flats Nuclear Disposal Superfund site (hereafter Maxey Flats site) accepted low level radioactive waste (LLW) from 1963 to 1977. The site comprised 46 unlined trenches used to dispose of both solid and solidified liquid wastes. An environmental monitoring programme that started in the 1970s indicated that contaminants were migrating from the waste via the shallow groundwater body at the site. Operations were stopped in 1977 and the site was stabilized, maintained and monitored. After additional evaluation and planning, the United States Environmental Protection Agency (US EPA) issued the Record of Decision Guidance, setting out the works necessary to enable site closure. This required the installation of an engineered cap, the construction of water management infrastructure, and a stewardship programme to maintain long term control over the site. Figure 9 presents a schematic of the engineered capping layers used to cover the waste trenches at the Maxey Flats site.

FIG. 8. Aerial view of the Southern Storage Area, Harwell, UK, during remediation work. Photograph courtesy of the United Kingdom Atomic Energy Agency and Magnox Ltd.

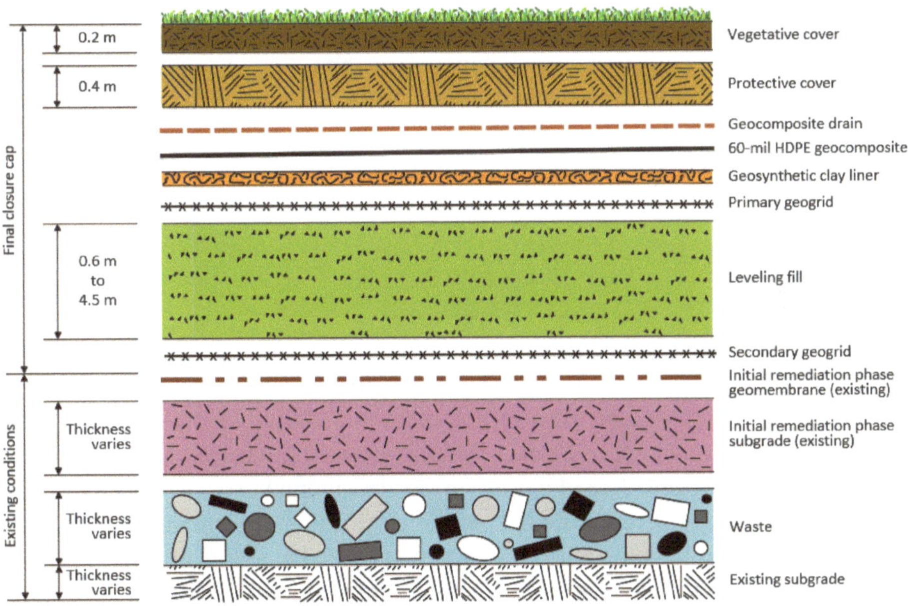

FIG. 9. Schematic of the engineered capping layers used to cover the waste trenches at the Maxey Flats site, USA. Reproduced from a figure courtesy of Commonwealth of Kentucky. HDPE — high density polyethylene.

FIG. 10. Aerial photograph of the Maxey Flats site, USA, showing the high density polyethylene geocomposite layer. Photograph courtesy of Commonwealth of Kentucky.

Figure 10 is an aerial photograph of the Maxey Flats site, showing the high density polyethylene geocomposite layer prior to the placement of the protective cover and vegetative cover. This site will be continually managed and monitored in perpetuity by the United States Department of Energy (US DOE) Office for Legacy Management [5]. Further information on the Maxey Flats site can be found in the case study presented in Appendix I.

2.5. REGULATORY CONSIDERATIONS

It is important that legacy trench sites are managed and remediated within a legal and regulatory framework that seeks to provide protection to human health and the environment from radiological and chemical risks and hazards. The legal and regulatory frameworks will promote an optimized and sustainable approach, allowing the consideration of environmental, social and economic factors in the selection of remedial activities. Further information on how to undertake sustainable remediation assessments can be found in ISO Standard 18504:2017 [6] and the UK's Sustainable Remediation Forum (SuRF-UK)

Framework[3] [7]. A graded approach to the regulatory requirements that is proportionate to the hazard and risk posed by the site is recommended [2].

In many Member States, nuclear safety is regulated separately from environmental safety. The impact of radiological and chemical contaminants may also be assessed under different regulatory frameworks. Hence, in this publication, the term 'regulatory requirements' is used to refer to both nuclear and environmental safety of the site. Often, more than one regulatory body may be responsible for the site and there may be more than one technical team from a regulatory body assessing and interacting with the site. It is therefore preferable to interact with all regulatory bodies to achieve agreement on the approach to managing and/or remediating the site.

Regulatory requirements will cover design, operation, decommissioning and/or remediation, and eventually closure of a site. For legacy trench sites, the focus is the remediation and/or closure of the site through achieving a site end state. Regulatory requirements from different regulatory frameworks need to fit together to provide the appropriate level of regulatory oversight throughout the lifetime of the site.

The regulatory requirements for the operation (and closure) of nuclear sites have developed considerably since the start of nuclear research in the 1940s and power generation in the 1950s. Early trench sites were either not regulated at all or not regulated to meet current regulatory requirements. This lack of regulatory oversight has compounded poor practices; for example, inappropriate disposal methods or inadequate record keeping. Since many waste trenches existed on operational nuclear sites, often permits or licences were issued for the operation of the whole site, rather than specifically of the trenches. Permits or licences generally focused on the operation of the site (and/or trench) and did not consider the closure of the site. In accident situations (e.g. accident at the ChNPP, Ukraine), action was needed immediately to deal with the large volumes of radiologically impacted wastes and hence there was no time to arrange permits or licences.

Over time, regulatory requirements have increased, although no country appears to have developed a specific set of regulations that purely focus on legacy trench sites. The approach has been to adapt or retrofit existing regimes to the regulation of legacy trench sites. These requirements consider the ongoing management of the site, as well as the evaluation and planning for the remediation and/or closure to ensure that the site does not pose a current or future risk to people or the environment. The implementation of new regulatory requirements

[3] SuRF-UK is part of the International Sustainable Remediation Alliance (IRSA). Among IRSA partners, the UK has taken the lead in developing guidance documents to support sustainable remediation practices.

has often been a driver to consider the site end state, including defining what is technically possible.

Depending on the country and the site context, some legacy trench sites are regulated as an existing exposure situation[4] (e.g. burials of waste from the accident at the ChNPP, Ukraine [9]) and some are regulated as a planned exposure situation[5] (e.g. LLWR, UK [10]). The situation can be further complicated when a legacy trench site is considered an existing exposure situation, but it is located within an operational site where the requirements for planned exposure situations apply. Hence, it is important that open dialogue is maintained between the operator and the regulatory body to determine the most effective way to regulate these sites and to achieve an optimized end state.

The key difference between existing exposure situations and planned exposure situations is application of the principles of optimization and of dose limits. As set out by the International Commission on Radiological Protection (ICRP) [11] and replicated in IAEA Safety Standards Series No. GSR Part 3, Radiation Protection and Safety of Radiation Sources: International Basic Safety Standards [8], a dose constraint is used to optimize protection and safety for planned exposure situations, and a reference level is used to optimize protection and safety for existing exposure situations. For a planned exposure situation, the ICRP and the IAEA specify a dose limit for a member of the public of 1 mSv per year. For an existing exposure situation, the ICRP specifies a reference level as a maximum of 20 mSv per year, and GSR Part 3 [8] further qualifies this by stating:

---

[4] GSR Part 3 [8] states:

"An *existing exposure situation* is a situation of exposure that already exists when a decision on the need for control needs to be taken.

"Existing exposure situations include exposure to natural background radiation that is amenable to control; exposure due to residual radioactive material that derives from past practices that were never subject to regulatory control or exposure due to residual radioactive material deriving from a nuclear or radiological emergency after an emergency has been declared to be ended."

[5] GSR Part 3 [8] defines a planned exposure situation as:

"[A] situation of exposure that arises from the planned operation of a source or from a planned activity that results in an exposure due to a source. Since provision for protection and safety can be made before embarking on the activity concerned, associated exposures and their probabilities of occurrence can be restricted from the outset. The primary means of controlling exposure in planned exposure situations is by good design of installations, equipment and operating procedures…In planned exposure situations, exposure at some level can be expected to occur. If exposure is not expected to occur with certainty, but could result from an accident or from an event or a sequence of events that may occur but is not certain to occur, this is referred to as 'potential exposure'."

> "Reference levels shall typically be expressed as an annual effective dose to the representative person in the range of 1–20 mSv or other corresponding risk quantity, the actual value depending on the feasibility of controlling the situation and on experience in managing similar situations in the past."

Recommendations on the application of the safety requirements for existing exposure situations are provided in GSG-15 [2]. In all exposure situations, the principles of justification and optimization need to be applied to ensure that doses are as low as is reasonably achievable, taking economic and societal factors into account.

2.6. INTERNATIONAL GUIDANCE

There are no international safety requirements or guides that are specifically written for legacy trench sites, although the management of historic wastes is recognized in many IAEA publications (e.g. Refs [12–15]).

For sites undergoing decommissioning, remediation and/or waste management activities, the IAEA safety fundamentals, relevant safety requirements and recommendations and the relevant ICRP recommendations apply (e.g. Refs [8, 16–20]). The IAEA safety standards can be used as a basis to develop a regulatory framework for the operation and closure of legacy trench sites.

3. WASTE INVENTORIES AND WASTE EMPLACEMENT

Legacy trench sites are often stand out from other contaminated sites because of the uncertainties arising from a complex contaminant source created by the disposed waste. This can be further complicated by poor or missing information regarding the waste inventory, waste form and waste location.

3.1. INVENTORY RECORDS

Inadequate, inaccurate, missing and/or contradictory information are common problems in inventory records for legacy trench sites. This leads to a high degree of uncertainty regarding the type and concentration of contaminants, the potential mobility of contaminants and the volumes of impacted ground and wastes.

Inventory records are an important data set for developing the conceptual site model and the source term for environmental modelling. Investing resources into identifying and evaluating inventory records can reduce the uncertainties in risk assessment and the planning of remedial activities. However, a balance is needed to ensure that resources are used effectively and efficiently to support all forms of information and data capture (e.g. non-intrusive and intrusive site investigation). It is only possible to obtain a limited amount of information from a poor record, and further review and cross-examination can only substantiate or improve the information to a certain extent. It is important to recognize that, regardless of the quality and comprehensiveness of the records, site characterization activities will be required to develop the conceptual site model.

Often, poor quality information is a result of practices that were considered appropriate at the time but do not meet current standards. Detailed spatial information may have not been considered necessary, as there was no expectation that the waste would be retrieved in the future. The information is often overly generalized and not sufficiently specific. For example, the location of an item may be described as "in trench 2" rather than a precise location, such as "at the base of trench 2, approximately 3 m below ground level and 4 m from the eastern end of the trench". Furthermore, the item may be described only as a 'drum', with no further description of the size, drum material or drum content. In some instances, the waste may have been placed in a completely different location.

Often, characterization data were composed of dose or total activity measurements, rather than data for specific radionuclides, and little or no information was provided regarding the chemical hazards of the waste. A possible common reason for this was that the primary use for the data was for worker safety, rather than supporting the disposability of the waste. The waste inventory data for waste disposal at the LFLS were recorded on 'pink cards' (Fig. 11) or a 'waste burial book' (Fig. 12). Only in some cases was the 'certificate' number in the waste burial book related to a pink card. Generally, little information was provided on the type of waste or on the inventory of radioactive or hazardous contaminants. Vague descriptions such as 'M.F.P.', standing for 'mixed fission products', were commonly used [21]. This type of record keeping can be seen at other legacy trench sites; for example, at Sellafield in the UK, Chalk River in Canada and several of the legacy trench sites in the USA.

There are also instances where inaccuracies and omissions have occurred. An example of this can be seen at the LFLS, where the summary records indicated disposal of either no tritium or small amounts of tritium. However, other records and site monitoring information indicate greater volumes and concentrations of tritium [21]. At the LFLS, 'detective work' has been necessary to support a

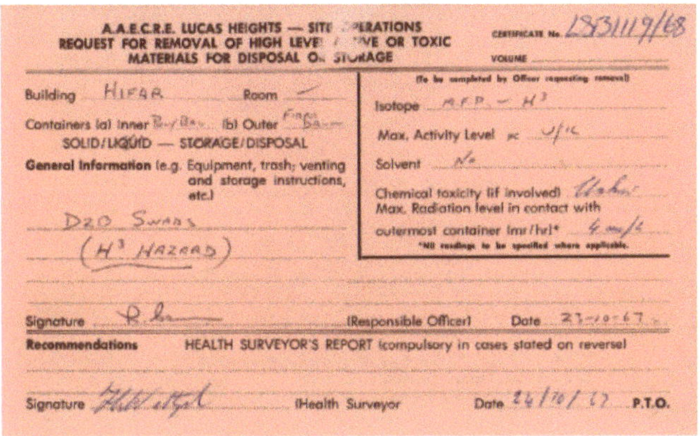
A.A.E.C.R.E. LUCAS HEIGHTS — SITE OPERATIONS
REQUEST FOR REMOVAL OF HIGH LEVEL ACTIVE OR TOXIC
MATERIALS FOR DISPOSAL OR STORAGE

CERTIFICATE No. 1853119/68
VOLUME

Building HIFAR Room —
Containers (a) Inner (b) Outer
SOLID/LIQUID — STORAGE/DISPOSAL
General Information (e.g. Equipment, trash; venting and storage instructions, etc.)
D2O SWABS
(H^3 HAZARD)

(To be completed by Officer requesting removal)
Isotope MFP — H^3
Max. Activity Level
Solvent No
Chemical toxicity (if involved)
Max. Radiation level in contact with outermost container (mr/hr)*
*Nil readings to be specified where applicable.

Signature (Responsible Officer) Date 23-10-67
Recommendations HEALTH SURVEYOR'S REPORT (compulsory in cases stated on reverse)
Signature (Health Surveyor Date 24/10/67 P.T.O.

FIG. 11. Pink card from the Little Forest Legacy Site, Australia. Photograph courtesy of the Australian Nuclear Science and Technology Organisation.

better understanding of the inventory. For example, from the pink card shown in Fig. 11, it has been possible to cross-reference the source of the tritium, which was HIFAR[6], with historic operational information on the tritium content of the reactor cooling circuit. From the waste burial book, the delay between waste generation and disposal — which is greater than two years, as shown in the example in Fig. 12 — attests to the likelihood that the waste had a much higher initial activity, which was allowed to decay prior to disposal. This theory is verified by other documentation.

Even with relatively well documented trenches, there may still be inaccurate information. For example, at the Solid Waste Storage Area of Oak Ridge National Laboratory, records showed that the waste casks were placed in a single layer. However, during the retrieval works, some casks were found to be stacked on top of each other. Further research of the disposal practices indicated that the approach to cask placement was evolving as waste disposal operations continued at the site [22].

In some cases, the waste records might not match the content of the trenches, because the waste may have been removed and this was not documented and/or the original waste record was not updated. In certain instances, the waste may have degraded beyond recognition because of the environmental conditions in the trenches.

---

[6] HIFAR: High Flux Australian Reactor.

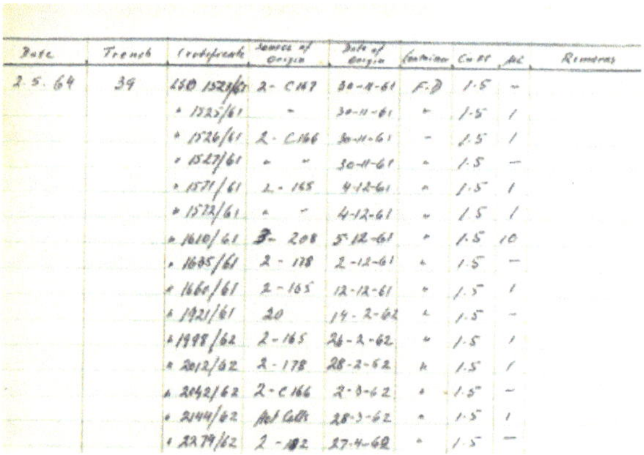

Date	Trench	Certificate	Source of Origin	Date of Origin	Container	Cu Ft		Remarks
2.5.64	39	150 152[illegible]/61	2-C147	30-11-61	F.D	1.5	–	
		" 1525/61	"	30-11-61	"	1.5	1	
		" 1526/61	2-C166	30-11-61	"	1.5	1	
		" 1527/61	" "	30-11-61	"	1.5	–	
		" 1571/61	2-165	4-12-61	"	1.5	1	
		" 1572/61	" "	4-12-61	"	1.5	1	
		" 1610/61	3-208	5-12-61	"	1.5	10	
		" 1685/61	2-178	2-12-61	"	1.5	–	
		" 1660/61	2-165	12-12-61	"	1.5	1	
		" 1921/61	20	14-2-62	"	1.5	–	
		" 1998/62	2-165	26-2-62	"	1.5	1	
		" 2012/62	2-178	28-2-62	"	1.5	1	
		" 2042/62	2-C166	2-3-62	"	1.5	–	
		" 2144/62	[illegible]	28-3-62	"	1.5	1	
		" 2279/62	2-182	27-4-62	"	1.5	–	

FIG. 12. Trench waste disposals reported in a waste burial book at the Little Forest Legacy Site, Australia. Photograph courtesy of the Australian Nuclear Science and Technology Organisation.

3.1.1. Loss of information

Information can be lost in several ways, including physical destruction of records due to poor storage conditions or accidents; loss of records during organizational changes; and loss of corporate memory as people retire or leave the organization.

Deterioration of paper files is a common problem in archiving if the storage conditions are not appropriate (e.g. low humidity). In addition, accidents such as fires and malignant actions can also lead to the loss of archived material.

It may be assumed that the archives of a government or its agencies are robust compared with those of private organizations. However, the dissolution of the USSR and the subsequent loss of records pertaining to the RADON type sites highlights that no organization is exempt from the risk of information loss.

There is a particular risk for information loss during the sale, transfer of responsibility or merging of organizations. These risks are not only related to paper and electronic files but also the loss of tacit information as the workforce changes. References [23, 24] describe good practices and lessons learned that can assist in preventing future information loss.

3.2. TYPES OF WASTE

Typically, early disposal practices were simple, and were applied without waste acceptance criteria[7] or a safety case. As a result, the wastes emplaced at legacy trench sites are often heterogeneous in type, form and contaminant. The waste can comprise both radioactive and non-radioactive wastes; the latter including hazardous materials such as explosives, chemical warfare reagents, ammunition, heavy and toxic metals, asbestos and organic solvents. Wastes can also include LLW, intermediate level radioactive waste (ILW) and sometimes high level radioactive waste (HLW)[8]. Typical radioactive contaminants include caesium-137, strontium-90, cobalt-60, americium-241, tritium and radioisotopes of plutonium, uranium, thorium. The wastes can be in an unconditioned and/or unpackaged form, which may facilitate the degradation of the wastes at a greater rate than if preconditioning, conditioning and treatment had been used.

Waste disposal practices at many legacy trench sites have resulted in the co-disposal of a diverse range of solid wastes, including the following:

- Wastes from laboratories — for example, glassware, rubber gloves, electronic instruments, ceramics and metal equipment;
- Wastes from demolition or environmental cleanup, including brick and concrete rubble, steel reinforcement and soil;
- Wastes with high organic content including paper, contaminated clothing and rags, wooden laboratory apparatus, office furniture and, in some cases, human sewage and animal remains;
- Higher activity wastes[9] including resins, chemical sludge, spent filter cartridges collected from waste treatment facilities and maintenance work;
- Disused sealed radioactive sources from medical or industrial applications.

---

[7] Waste acceptance criteria are quantitative or qualitative criteria, specified by the regulatory body, or specified by an operator and approved by the regulatory body, for the waste form and waste package to be accepted by the operator of a waste management facility [25].

[8] LLW is radioactive waste that is above clearance levels, but with limited amounts of long lived radionuclides [25]. ILW is radioactive waste that, because of its content of long lived radionuclides, requires a greater degree of containment and isolation than that provided by near surface disposal [25]. HLW is the radioactive liquid containing most of the fission products and actinides present in spent fuel, which forms the residue from the first solvent extraction cycle in reprocessing, and some of the associated waste streams; this material following solidification; spent fuel (if it is declared as waste); or any other waste with similar radiological characteristics [25].

[9] For the purpose of this publication, the term 'higher activity waste' is used to include both ILW and HLW.

At some legacy trench sites, liquid wastes were poured directly into trenches, or contained in drums or solidified prior to disposal. Solidification often occurred by mixing with sand or cement, and then pouring the mixture into a concrete box or drum. Liquid wastes may have been generated from process or maintenance activities and, in some cases, organic liquids were burned inside or in the vicinity of trenches.

As a quick disposal option, trenches were also used for the disposal of wastes from nuclear incidents or accidents. Again, waste types varied and could include personal protective equipment (e.g. suits, rubber gloves), contaminated machinery and vehicles, and soil and contaminated vegetation. Typically, the inventory of these trenches was undocumented, since there would have been little time to document the content during an emergency situation.

The type of waste and contaminant varies from site to site and reflects the site operations, which may have also evolved over time. Even within a single waste trench, the inventory can be heterogeneous, with pockets of waste that are distinct, either physically or chemically, from neighbouring waste. This heterogeneity makes environmental modelling challenging and can significantly affect estimations for waste retrievals or remediation, as well as present complex safety hazards during site works.

Experience has shown that components of the waste have interacted and increased the mobility of radioactive contaminants. An example of this effect was observed at the Maxey Flats site, USA, where the mobility of plutonium was increased by the presence of strong complexing agents such as ethylene diamine tetraacetic acid[10] [26]. At the LFLS in Australia, research has identified an ongoing tri-butyl phosphate release from wastes in the trenches. The presence and solubility of tri-butyl phosphate could provide a mechanism for radionuclide mobilization, which may have long term implications for the site [27].

In some cases, wastes containing non-radioactive contaminants may present a greater risk to people and the environment than the radioactive contaminants in the waste do. Hence, it is important not to ignore non-radioactive contaminants or consider them less important. Like the radioactive inventory, the non-radioactive inventory will reflect the wastes generated from the various activities at a site. For example, where demolition has occurred, wastes may include light fixtures and electrical cables, which can contain polychlorinated biphenyls[11], or lead shielding could be present if it was contaminated during an experiment and hence disposal was necessary.

---

[10] More commonly known as EDTA.

[11] More commonly known as PCBs.

The Southern Storage Area in Harwell, UK, is an example of a legacy trench site where the chemical hazards posed a greater risk than the radioactive hazards. Beryllium[12] was the main contaminant of concern, but mercury and other heavy metals were also present [28]. Beryllium is also a key contaminant in the legacy trenches at the LFLS, Australia [21].

3.3. LOCATION AND CONSTRUCTION OF LEGACY TRENCH SITES

Often, the construction of a legacy trench site was in response to an operational need or, in some cases, to an incident. Frequently, the location or construction of the waste trenches was not driven by environmental suitability, but by operational needs. In some instances, the proximity of the disposal site to the location where the waste was generated was considered an advantage. In other cases, moving waste away from operations ensured that waste management activities would not impact operations. Moreover, by selecting a location that was difficult or unlikely to be developed, it was assumed that the site could be vacated once disposal activities were complete. If the trenches were to support operations from several operational facilities, a location may have been selected that was easily accessible to all facilities.

Construction was often simple, with little or no consideration given to isolating the waste from the surrounding soils and groundwater. As disposal practices evolved, trenches included engineered barriers; however, the quality of their construction often failed to keep the contaminants within the structure.

Typically, trench locations were not marked or, if field markers were used, they may have been moved or removed by subsequent caretakers of the site. Where drawings were created, they were typically hand-drawn sketches and not surveyed plans.

Settlement can be an issue even in conventional landfills, and is a particularly important process where waste has been tumble-tipped and not compacted, as is often the case at legacy trench sites. Voids can be caused by the following two main processes:

(a) The wastes settling (consolidation and compression) over time owing to the weight of overlying wastes and/or the movement of water and gas

[12] Beryllium is a lightweight, but toxic, metal that has excellent neutron absorbing and moderating properties. It is used as a neutron source or reflector and is present in radiation windows and instrumentation. It is therefore commonly found in waste streams associated with nuclear research facilities.

through the waste. This may include the collapse of bulky items with voids (e.g. glove boxes, empty containers) (Fig. 13).

(b) The degradation of items such as wooden work benches, vegetation and clothing or the corrosion of metal items.

Figure 13 shows the collapse of a trench cover that has revealed a waste container at the '3rd Stage of the Chornobyl nuclear power plant' burial site. This waste burial site contains radioactive wastes from the cleanup of the ChNPP sarcophagus site, which took place in 1986–1987. The photograph in Fig. 13 was taken in 2015. The void has subsequently been filled.

3.4. WASTE DISPOSITION

Typically, bulk wastes were tumble-tipped into trenches with no prior sorting or packaging (Fig. 14). In general, higher activity wastes were emplaced in engineered trenches. In some instances, loose wastes were packed into drums, which were stacked neatly to conserve space (Fig. 2). Depending on the environment, the use of drums provided an additional level of containment within the trench system. In contrast to drummed waste, often liquids were poured directly into the ground, pits or trenches. In some instances, trenches contained surface water or intercepted the shallow groundwater or perched water bodies. This increased the volume of contaminated liquids in the disposal trench (Fig. 15).

FIG. 13. Void created by the collapse of a trench cover at the '3rd Stage of Chornobyl nuclear power plant' burial site, Ukraine. Photograph courtesy of D. Bugai, Institute of Geological Sciences, Kiev.

FIG. 14. Historic waste disposed at the RADON type facility at Tammiku, Estonia. Photograph courtesy of Estonian Waste Management Organisation (A.L.A.R.A Ltd).

FIG. 15. Liquid accumulated in an unlined waste trench at the Chalk River site c. 1953. Photograph courtesy of Atomic Energy of Canada Ltd.

3.5. WASTE VOLUMES

Generally, the volume of waste in trenches decreases at a legacy trench site as degradation occurs in the unconditioned waste. However, as contaminants migrate to the surrounding environment, the overall volume of waste and impacted materials that need management will be larger than the original volume of waste.

For non-soil wastes, conditioning could subsequently lead to either a smaller or a larger volume of waste to be managed. Processes such as size reduction and compaction can reduce waste volumes, whereas grouting wastes will increase the waste volume.

4. SITE EVALUATION

This section provides information and guidance on the characterization of legacy trench sites to underpin management and/or remediation decisions and activities.

When considering corrective action, the following types of information are necessary to enable site evaluation and underpin decisions:

- Baseline site characterization data, if available;
- The types of waste and associated contaminants;
- Information on changes over time to the site or trench conditions;
- Knowledge of the performance of any engineered barriers that may have been utilized in the trenches;
- Assessments on the extent of water ingress and egress;
- Environmental monitoring data for all relevant media (e.g. soils, water, air, airborne dust, vegetation).

The practice of site evaluation, which includes site characterization, generally follows a process consisting of the following types of activity:

- Desk study;
- Walkover survey;
- The production of an initial conceptual site model (CSM);
- Non-intrusive site characterization;
- Intrusive site characterization;
- Revision of the CSM.

The site characterization and the revision of the CSM often constitute an iterative process; each phase of works provides additional information, which can be used to target uncertainties in the understanding of the environmental processes at the site.

The desk study and walkover survey allow the collation of all readily available information and are used to produce the initial CSM and inform the site investigation works. The CSM records what is understood about the site and its surroundings and what is still unknown, and articulates the potential risks posed by the site to people and the environment.

The CSM sets out the source–pathway–receptor linkages and the basis for safety assessment and environmental modelling. Figure 16 provides an example of two potential pollutant linkages at a legacy trench site, where the main source of contamination is the waste in the trenches. Pathways include leaching of contaminants from the waste and movement within groundwater bodies or surface water through bath-tubbing, airborne dust, bioaccumulation, human intrusion into the trenches, extraction of contaminant impacted groundwater, and waste or soil being brought to the surface by burrowing animals. Typical receptors include protected water bodies, as well as livestock and people living near the site.

Often, the CSM is a simple flow diagram or pictorial representation of the system that illustrates the pollutant linkages. This diagram is then augmented with other detailed data sets (e.g. hydrogeological groundwater models, monitoring data). The CSM is a 'living' document and will be continually updated as further site characterization and assessment are completed. The initial version is used to design the characterization programme, and later versions underpin contaminant modelling and support decisions regarding management and/or remediation options. It is important that all the knowledge and information used to develop the CSM are maintained to ensure the integrity of decisions regarding the site.

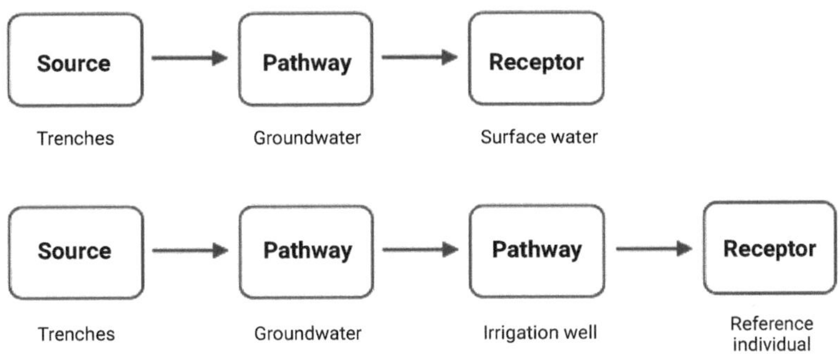

FIG. 16. Schematic of potential pollutant linkages at a legacy trench site.

Further guidance on the development of a CSM can be found in ISO Standard 21365:2019 [29]. Section 4.4 provides further guidance on developing the CSM for the purpose of environmental modelling and safety assessment.

4.1. RECONSTRUCTING THE HISTORY OF A LEGACY SITE

Management and, where necessary, remediation of any contaminated land site require the collection of knowledge and information to inform and underpin decision making. An important part of this process, especially for a legacy trench site is reconstructing the site history. The robustness of the decision will rely on the quality of the underpinning information and knowledge upon which it was based. Consideration needs to be given to three knowledge and information management areas:

(a) Desk based studies intended to recover or reconstruct the history of the site (this section);
(b) Site characterization to gather new information to assess the impact of the site on people and the environment (Section 4.3);
(c) Management and preservation of records to document actions and underpin future decisions (Section 8).

A unique challenge for legacy trench sites is the extent of work that is necessary to understand their history. Knowledge and information are needed on the location and setting of the trench, the construction of the trench, the disposed waste (i.e. the source) and trench operational decisions.

Gathering and maintaining knowledge and information can be particularly difficult where there is no responsible organization identified for the site. It is also likely that knowledge and information will be less available for those sites that were operated with limited or no regulatory oversight.

Further information on undertaking a desk study can be found in Refs [2, 30, 31].

4.1.1. Assembling existing information on the site history and waste disposals

In most cases, basic information on the trenches and the surrounding environment does exist. However, the availability of detailed information may be scarce and/or unreliable. Information needs to be gathered on all potential hazards at the site (e.g. radiological, chemical, ground stability).

Information sources for the desk based study include the following:

— Excavation or engineering plans;
— Waste disposal records;
— Records providing information on the waste producer's activities (e.g. process, procedures);
— Existing site characterization and/or monitoring data;
— Photographic records, for example, aerial photographs or photographs taken during site operations;
— Topographical survey data and mapping information;
— Documents recording past decisions;
— Historic and current regulatory requirements and/or agreements;
— Knowledge gathered by interviewing existing or former personnel;
— Operation permits, licences or authorizations.

If the legacy trench site is located close to where the wastes originated, the past activities at the site are likely to reflect the types of waste disposed of in the trenches. If records are limited or non-existent, it will be beneficial to collect anecdotal evidence by interviewing existing or former site personnel. Knowledge gathered through the interview process needs to be documented for future use.

As an example, at the LLWR site, UK, waste was historically tumble-tipped into seven trenches (Fig. 1). A waste inventory was developed for the trenches using disposal records for the site and the consigning operator, historic photographs and drawings, knowledge recorded from current and past employees at the site, and environmental monitoring and modelling. On the basis of these records, the operator was able to develop maps of the trenches showing radionuclide distribution. Although it was recognized that some uncertainty remained, the radionuclide distribution map and other forms of evidence (e.g. contaminant modelling) were used to underpin the Environmental Safety Case and the options assessment that demonstrated that the disposed waste can remain in the trenches [10].

Reference [32] provides guidance on the retrieval and restoration of waste inventory information. This document sets out a stepwise and prioritized approach for investigating and retrieving information, which includes the following:

— Assessment of the reliability of the inventory records;
— Implementation of data retrieval and the assessment methodology;
— Data verification and validation.

4.2. WALKOVER SURVEY

A walkover survey[13] is a routine part of the site evaluation phase and further guidance can be found in ISO Standard 18400-202:2018 [31]. When carrying out a walkover survey on a known or potential legacy trench site, there are specific features that need to be identified or that may indicate that trenches are present.

Apart from looking for markers that may delineate a trench, areas of ground subsidence can indicate that the ground may have been disturbed during the formation and placement of waste in a trench or that voids have subsequently formed within the backfilled waste or cover material.

Another key feature to observe is whether there is any pooling or seepage of surface water in or close to the trench area. This may suggest bath-tubbing, which can bring contaminants to the ground surface around the trenches. Ideally, the site needs to be visited after a prolonged rain event to observe whether this phenomenon is occurring.

Walkover surveys can also be supported using aerial surveying technology, including drones and historic aerial photographs. Viewing a site from an aerial position can often identify features that cannot be seen from the ground. For example, it may be possible to delineate the position of the trenches from large area ground disturbances, evidence of subsidence, and variations in vegetation.

4.3. SITE INVESTIGATION

As highlighted in Section 4.1, information about a legacy trench site may be inadequate or missing. Site investigation supports several aspects of managing a legacy trench site: it provides information to support modelling and assessment to ascertain whether the site poses a risk to people and/or the environment; it is necessary to support remedial design and verification; and monitoring may be needed before and after remediation. Site investigation helps to fill in the gaps in site information and reduces uncertainty. Therefore, it is important that good quality data are available to underpin decision making and gain stakeholder support.

There are two aspects to consider when investigating a legacy trench site: one is characterizing the trenches and their content and the other is characterizing the surrounding ground and water bodies. If remediation is implemented, characterization of the wider site will aid the determination of baseline conditions that can be used in the development of remedial targets and support verification of the remedial actions.

[13] Also known as a site reconnaissance.

This section provides information on non-intrusive techniques used to understand the location, extent and content of the trenches and site investigation techniques to characterize the environment around the trenches. A brief discussion is provided in Section 4.3.2 on the characterization of trench waste whilst it remains in situ. Waste characterization following retrieval and the approach to predisposal activities are not discussed in this publication; waste characterization is described, for example, in Refs [3, 13, 33–36].

Site characterization of contaminated land sites is a mature subject widely discussed in the literature (for example, see Refs [31, 37–43]). However, the following features of legacy trench sites make them difficult to characterize:

— Heterogeneity of the waste in the trenches;
— Mix of radiological, chemical and physical hazards;
— Presence of discrete items or packages of concern;
— Physical difficulty of investigation.

It is important that site characterization is not viewed as a one-off event. Several site investigations may be needed, which build on the information gathered in early investigations. A successful characterization programme is often one that combines a range of different techniques (non-intrusive and intrusive surveys). Following any site investigation works, the CSM needs to be updated to support further site characterization and/or remedial design. Site characterization may also be used to validate the success of the adopted remedial actions and support any future work, including additional remediation, if required.

Site characterization is an expensive activity and needs to be carefully planned and phased to ensure that the correct information is collected to support the characterization approach and to address data gaps. ISO Standard 18400-101:2017 [44] provides guidance on the principles and key elements of the sampling plan. The US EPA has developed the Data Quality Objectives process [45] to support systematic planning for site characterization. The Data Quality Objectives process is a series of logical steps used to plan for the resource effective acquisition of data of sufficient quality and quantity to support the goals of the study.

4.3.1. Non-intrusive site investigation

Non-intrusive techniques, such as radiological or geophysical surveys, need to be considered before undertaking a programme of intrusive site investigation. There are several benefits to this approach:

— Non-intrusive techniques can be used to target further site investigation activities.
— Non-intrusive techniques, including radiological surveys, are generally quicker, can generate large data sets, are less expensive and safer to implement than intrusive techniques.
— In many instances, the data gathered can be viewed in real time, which will inform the next phase of works.
— The use of non-intrusive techniques minimizes the chance of disturbing buried containers, which may lead to the release of hazardous materials into the air or groundwater.
— Non-intrusive techniques do not create further waste materials.

Several non-intrusive techniques are particularly useful for investigating legacy trenches, and they are generally more effective when used together rather than in isolation. The two most commonly used approaches to non-intrusive surveys are radiological and geophysical surveys.

4.3.1.1. Radiological surveys

Radiological surveys use portable detection equipment that responds rapidly to radionuclide contamination. Techniques such as gamma surveys can be used to gain an understanding of near surface radioactive contamination. While many radionuclides associated with radioactive waste are gamma emitters, alpha and beta emitting isotopes cannot be detected and require more complex measurement techniques.

There are two approaches used for near surface soil characterization: direct (point) measurements and scanning systems. Direct measurement systems can provide absolute values for certain parameters or identify additional information on the range of radionuclides present. These systems are generally bulky and are used for stationary measurement for a set period. Scanning systems use mobile equipment that can either be carried or mounted on a trolley or small vehicle. These systems typically use sodium iodide detectors and focus on the total activity of gamma emitting radionuclides. Therefore, large areas can be scanned for a relatively low cost. The output survey provides rapid information about the relative levels of radioactivity in the near surface soils across the site. This information can be used to focus intrusive sampling.

At legacy trench sites, the collected data may be correlated to cross-contamination during disposal activities, subsequent disturbance of the trenches, transport by infiltration water or groundwater, or bath-tubbing effects. Reference [46] provides a review of methodologies for in situ characterization using nuclear spectrometry techniques. References [41, 42] provide overarching guidance on radiological surveys.

4.3.1.2. Geophysical surveys

Geophysical techniques can be used to measure near surface structures or changes in ground chemistry as well as lithology. Geophysical techniques can detect the presence of large buried objects, as well as the presence of voids and subsurface water bodies. These techniques are often used at legacy trench sites to determine the location and extent of the trenches. This can be challenging on an operating site where the trenches have been excavated into disturbed ground or filled areas, since it may be difficult to differentiate the instrument signal of the trenches.

Table 2 summarizes commonly used geophysical techniques and their specific application to legacy trench sites. Overarching guidance on geophysical surveys can be found in Ref. [41] and detailed information can be found in Refs [30, 47–49].

Surface and down-hole geophysical surveys were used to investigate the location of the Windscale Trenches at the Sellafield site, UK. The six unlined trenches cover an area of approximately 7000 m^2 and were the main on-site disposal for solid wastes in the 1950s. Although the records were poorly kept, the waste inventory is known to include tritium, other fission products, actinides, solvents and asbestos. The geophysical surveys provided some additional information regarding the location and content of the trenches but were limited because it was difficult to insert the probes into the dense and compacted ground in the trench area [50].

A series of site characterization activities have been undertaken at the LFLS, Australia, since the cessation of disposals in 1968. Non-intrusive characterization, including ground penetrating radar, electromagnetic and electrical resistivity tomography surveys, have been used to investigate and support understanding of the trenches [51, 52]. The ground penetrating radar provided limited information, and it was difficult to delineate the edge of the trench from the undisturbed ground. The electrical resistivity surveys enabled delineation of the trench structure and surrounding geology and provided information about the trench content and the groundwater levels in the trenches. The electromagnetic survey

TABLE 2. EXAMPLES OF COMMONLY USED GEOPHYSICAL TECHNIQUES AND THEIR APPLICATION TO LEGACY TRENCH SITES

Technique	Application
Ground penetrating radar	This technique is based on the transmission of electromagnetic energy pulses into the ground and the measurement of the amplitude and travel time of the signals as they return. This is a useful technique for identifying disturbed ground, subsurface features or structures such as foundations and concrete slab thickness, voids and metal objects such as drums. In the case of legacy trenches, it helps to delineate the trench boundaries and highlight any specific waste containers close to the surface. Ground penetrating radar performs best on dry sandy soils and poorly on wet clays.
Electrical resistivity profiling	This technique requires the insertion of an array of electrodes into the ground surface, through which an electrical current is passed. By measuring the changes in electrical potential between the electrodes, buried metallic objects can be identified, and changes in ground conductivity related to groundwater chemistry or flow direction may be determined. The technique is useful where above ground structures impede the use of techniques that require movement of the ground surface.
Electromagnetic surveying (also known as electromagnetic ground conductivity survey)	This technique uses electrometric induction to determine the electrical properties of buried items and it can be utilized to detect metallic objects, subsurface voids and subsurface features or structures. The exact shape, size and contents of such objects cannot be determined through this approach. However, the general locations of these objects can be identified and, if isolated, their size can be estimated. This is a non-contact technique where the equipment is moved over the ground surface.
Seismic refraction	This technique is based on the refraction of seismic energy at subsurface interfaces. The equipment comprises an array of geophones and a seismic source. As the seismic waves encounter an interface between different rock or soil layers, a portion of the seismic energy is reflected. This enables the creation of a seismic velocity cross-section, which enables subsurface structures, such as trenches, to be identified.
Microgravity survey	This technique measures extremely small variations in the earth's gravitational field caused by the different densities of materials or voids under the surface. It can be used to determine density differences between the trenches and their surrounding material.

demonstrated good correlation with the inventory records, which identified the disposal of drums in specific trenches. The 2017 surveys were also used to confirm that the location of the proposed pilot trench experiment was an area of undisturbed ground.

4.3.2. Intrusive site investigation

The desk study, CSM and non-intrusive surveys provide an initial understanding of the site. These site evaluation activities, especially the CSM, will inform the design of the intrusive characterization programme. Intrusive site investigation can primarily be used to gather information regarding the geological and hydrogeological setting of the site. Intrusive investigation can also be used to investigate the contents of the trench; however, a decision to disturb the trenches needs to be justified against any potential increases of risks to people and the environment and the generation of waste.

The information collected can be used to underpin modelling and remedial design efforts. Understanding the geological and hydrogeological regime is particularly important in the selection of remedial options that isolate or stabilize the trenches in situ.

Undertaking an intrusive investigation requires thorough planning to ensure that all aspects of health and safety and protection of the environment have been addressed. Additional precautions, such as containment tents and hazmat suits, may be necessary where there are higher uncertainties associated with investigations close to or into the trenches. Intrusive works need to be discussed with the regulatory authorities and may need approval prior to starting.

4.3.2.1. Intrusive investigations around the trenches and across the site

If there is high confidence in the results obtained from non-intrusive techniques, intrusive techniques can be used next to trenches to confirm the extent of the trenches and to understand the surrounding ground and groundwater bodies. Groundwater wells adjacent to the trenches can provide useful data on the movement of contaminants from the trenches. A wider intrusive investigation of the site may also be beneficial to confirm the geological and hydrogeological context of the trenches. In particular, such an investigation will be necessary to develop an understanding of the groundwater bodies and their movement across the site. Where the trenches are suspected to contain gas-generating materials (e.g. high organic content material), wells will also be needed to monitor ground gas evolution.

Table 3 provides a summary of the commonly used intrusive site investigation techniques. Further detailed information on intrusive site investigation techniques can be found in ISO Standard 18400-102:2017 [53, 54].

Cone penetration testing is another useful technique for identifying disturbed ground, conducting readings and taking samples. It is normally utilized for assessing the different geotechnical properties in soils but can be used for other purposes. In the case of legacy trenches, it can help to delineate the trench boundaries (both vertically and laterally), highlight any waste containers and take samples of the trench material. Cone penetration testing works through pushing a cone into the ground from a static fixed vehicle. If the cone cannot successfully push through or beyond any solid object encountered, then a new sampling location will need to be found. Different cones can be vertically advanced, each with a slightly different functionality, such as the following:

- A groundwater sampling cone can collect groundwater or leachate samples from below the base of the trenches. The samples taken will then need to be analysed in a mobile or off-site laboratory for radioactive and non-radioactive contaminants.
- A gamma cone can identify gamma emitting materials and help to produce a vertical profile of the trench contents.
- A resistance cone measures resistance to pressure and therefore can provide information on the different types of material that it passes through.
- A conductivity cone can detect the presence of water and, in some instances, additionally provide an assessment of the level of contamination in the soil.
- A video cone can, in certain situations, provide a visual view of the trenches as the cone is advanced.

An important point to note is that intrusive investigations are likely to result in material being brought to the ground surface. This material, if contaminated, may subsequently be classified as radioactive waste or other hazardous waste and would need to be managed accordingly. The potential production of waste and its subsequent management need to be considered when choosing the site characterization approach. The characterization process needs to be optimized so that as little waste as possible is produced.

There have been a series of non-intrusive and intrusive site investigations undertaken at the LLWR site, UK, dating back to 1939, when the site was developed as a Royal Ordnance Facility [56]. Since 1939, approximately 650 boreholes have been drilled either on or in close proximity to the site. These boreholes were drilled for several reasons, including for geotechnical investigations and as long term groundwater monitoring points. The boreholes that were not installed as monitoring wells were subsequently backfilled with the drilling arisings and/or

TABLE 3. SUMMARY OF INTRUSIVE SITE INVESTIGATION TECHNIQUES

Technique	Comments	Advantages	Disadvantages
Trial pits and trenches Depth: 0–6 mbgl[a]	Excavated by hand, vacuum excavation or machine excavator.	— Simple and relatively quick — Allows subsurface geology to be viewed in three dimensions — Easy to obtain samples (discrete and bulk) — Geological strata can be photographed and logged — Water strikes and some forms of contamination can be observed	— Depth achieved depends on geology and machinery used — Larger working area required compared with other techniques — Increased safety issues (e.g. falls, entrapment) compared with other techniques — Trench stability can be an issue, particularly below the water table — Can generate more waste compared with other techniques
Hand auger Depth: 0–2 mbgl	Many designs available for different soil types. Some equipment allows the collection of core samples.	— Simple and relatively cheap — Portable and therefore beneficial for locations where access is restricted	— Ease of use is limited by the ground conditions and/or soil type — Depths are limited, particularly if obstructions (e.g. stones) are present — Collection of only small sample volumes — Cross-contamination can occur along the bore unless plastic casing is used

TABLE 3. SUMMARY OF INTRUSIVE SITE INVESTIGATION TECHNIQUES (cont.)

Technique	Comments	Advantages	Disadvantages
Rotary drilling using solid stem auger Depth: 0–20 mbgl	Use of a power driven auger to create open-hole bores. Typically, lubrication is needed (air, water or drilling mud) to keep the cutting head cool. Primarily used to advance a borehole. Used to install gas or water monitoring wells.	— Can drive through hard materials such as boulders — Depending on the bore depth and geology, the hole will remain open after the auger is removed — Samples recovered using a wire line, double tube or triple tube (includes liner) barrels using rods	— Can be difficult to observe fluid stratigraphic features (e.g. gravels to a soft sandstone)
Rotary drilling using hollow stem auger[b] Depth: 0–20 mbgl	Use of a power driven auger to create open hole bores. Typically, lubrication (air, water or drilling mud) is needed to keep the cutting head cool. Samples can be collected via the hollow central shaft. Used to install gas or water monitoring wells.	— Samples can be collected through a hollow stem — Borehole is fully cased, which minimizes risk of cross-contamination — Good recovery of coarse samples (e.g. gravels)	— Difficult to measure groundwater strikes, especially if water is used during drilling

TABLE 3. SUMMARY OF INTRUSIVE SITE INVESTIGATION TECHNIQUES (cont.)

Technique	Comments	Advantages	Disadvantages
Window or windowless sampling Depth: Handheld: 0–6 mbgl Rig: 0–30 mbgl	Cylindrical steel tubes are driven into the ground using a hydraulic hammer. The hammer can be mounted on a rig or a portable jackhammer. Once the tubes are withdrawn, the core can be examined either through the window or as an intact core using a disposable plastic liner in a windowless tube. Used to install gas or water monitoring wells.	— Allows collection of intact samples — Use of plastic liner supports collection of volatile contaminants — Small rig can be used on small sites — Handheld suitable for areas with restricted access — Use of plastic liner and/or casing to minimize cross-contamination and ensure that the bore remains open	— Cannot drill through hard rock or concrete unless the rig has dual rotary and percussive ability — Difficult to measure groundwater strikes, especially if water is used during drilling
Cable percussion (shell and auger) drilling Depth: 0.5–30 mbgl	A cutting tool creates the bore via gravity percussion. The cutting tool is winched into position on a tripod derrick by a diesel engine. Used to install gas or water monitoring wells.	— Most suited to cohesive (e.g. clay) or non-cohesive (e.g. sand and gravels) — Hydrogeological information[c] can be gathered throughout the geological sequence [55] — Large bore can be created (up to 300 mm) — Casing used to minimize cross-contamination and ensure that the bore remains open — Undisturbed and disturbed samples can be collected	— Typically, more time consuming and messy than other methods — Difficult to measure groundwater strikes, especially if water is used during drilling — Relatively large working area required

TABLE 3. SUMMARY OF INTRUSIVE SITE INVESTIGATION TECHNIQUES (cont.)

Technique	Comments	Advantages	Disadvantages
Sonic/rotasonic drilling Depth: 0–20 mbgl/ 0–100 mbgl	Use of high frequency energy to displace and shear soils. Used to install gas or water monitoring wells. Rotasonic combines sonic and rotary drilling methods in one rig	— Enables collection of undisturbed samples and complete cores in most ground conditions — Faster drilling compared with other methods — Can be used on all soil types and concrete and hard rocks — Can be used without a drilling flush — Casing used to minimize cross-contamination and ensure that the bore remains open	— Difficult to measure groundwater strikes, especially if water is used during drilling

[a] mbgl: metres below ground level.

[b] Also described as continuous flight auger with hollow central shaft.

[c] Tests include rising head tests, falling head tests, standard penetration tests, particle size distribution, bulk density tests, moisture content tests and hydrogeochemical sampling.

bentonite clay. The bentonite clay was used as a 'plug' between geological units to prevent the creation of a preferential pathway for contaminant movement. LLWR also has procedures in place for decommissioning boreholes in line with regulatory guidance to prevent the creation of potential pollutant pathways. It is recognized that the more recently drilled boreholes provide a higher quality of information compared with previously drilled boreholes. These improvements reflect the advances in drilling techniques, borehole logging methodologies and the data quality objectives of the investigation programme. However, the data from the older boreholes are still recognized as a valuable resource.

4.3.2.2. Intrusive investigation into the trenches

Undertaking intrusive investigation into the trenches requires careful consideration. The decision needs to be justified with a clear explanation to the need and/or expected benefits, the uncertainties associated with undertaking the work, and a plan on how potential risks will be managed.

Characterization of heterogeneous trench contents presents its own unique challenges. Unlike contaminated soils, legacy waste trenches are more likely to be heterogeneous with discrete objects. It is therefore difficult to use intrusive investigation to collect a representative sample of the trench contents, and it is likely that only qualitative information on the materials and their hazardous properties can be collected. Realistically, it may only be possible to characterize materials from a legacy trench if there is a decision to retrieve the wastes from the trenches.

A key consideration is whether the potential health and safety risks to workers are justified compared with the potential benefits from gathering additional information from the legacy trenches. Radioactive and hazardous materials could be accidently released into the atmosphere and/or workers might be exposed. It is likely that discussions with the regulating authorities will be necessary and the proposed works will need to demonstrate that risks to the public and workers are as low as reasonably achievable (ALARA)[14].

Another health and safety related consideration is whether the intrusion into the legacy trench may compromise a waste package or container. This could lead to the release of a hazardous material into the trench and potentially into the environment via a groundwater or vapour pathway.

It is likely that a decision to intrude into a legacy trench has been taken as a pre-activity to underpin a decision to retrieve the waste. The aim of this

---

[14] For further information on the ALARA principle, see IAEA Safety Standards Series No. SF-1, Fundamental Safety Principles [16], and the definition for optimization (of protection and safety) in the IAEA Safety and Security Glossary [25].

investigation would be to collect information on the type and volume of material to inform predisposal and disposal activities.

There are many examples where an intrusive investigation requires excavation through a layer of capping material (see Table 1). However, in some instances, access to the waste may be comparatively simple because the legacy trenches have been covered with concrete slabs or metal sheets (e.g. RADON type facilities).

Window or windowless sampling, microdrilling and cone penetration testing are likely to be the most suitable techniques to conduct investigations inside the trench. Since these methods can create small boreholes, they will minimize the risk of negatively disturbing the material in the trench. In a few instances, it may be appropriate to remove an area of the trench cap via excavation, to enable a visual inspection of the waste and facilitate targeted sampling. This approach would require a greater degree of certainty regarding the trench content and measures to ensure protection of people and the environment.

At the UK's Harwell site, the Southern Storage Area comprised five unlined shallow trenches excavated into the chalky soil and capped with soil. The trenches included beryllium contaminated wastes originating from decommissioning of beryllium fabrication facilities, solvents, mercury and radioactively contaminated materials from site operations. A formal decision making process concluded that all the wastes needed to be retrieved in order to facilitate an unrestricted reuse of the site. An in-depth site characterization was needed to support the adopted remediation approach and the associated waste management activities. Following a desk study phase, an integrated non-intrusive and intrusive site investigation took place. The adopted approaches included walk-over surveys, trial pits, soil sampling, soil gas analysis, core sampling, geophysics, probe surveys and groundwater sampling. The intrusive investigations into the trenches were conducted using a double-layer contaminant tent, as well as personal protective equipment with respiratory protection. Cores were collected using a heavy duty window sampler, which provided undisturbed sample cores and minimized waste generation. Boreholes were drilled on a 1.5 m grid to between 3.5 m and 6.5 m below ground level [28].

4.3.3. Monitoring

Monitoring activities generally take place throughout the lifetime of a legacy trench site or any contaminated land site. Monitoring is used before, during and after remedial works and may be used in several ways, depending on the circumstances of the site, including the following examples:

— To provide baseline level data;
— To demonstrate that no further pollution is occurring;
— To validate the success of a remedial action (e.g. the continued decrease in contaminants measured in a groundwater plume after the installation of a cap or following removal of the waste);
— To support maintenance of a safety case to allow the continued presence of the legacy trenches.

Typically, groundwater monitoring is the focus of a monitoring regime at a legacy trench site. Ground gas monitoring may be necessary where the legacy trenches contain materials that can degrade to release gases (e.g. decomposition of materials such as wood, vegetation and clothing can generate methane and volatile organic compounds). Surface monitoring and/or sampling may also be important where bath-tubbing is known to occur, which can bring contaminants to the ground surface.

If the post-remediation monitoring programme shows that the remediation has been unsuccessful it may be necessary to undertake further site characterization to understand the extent of the problem and support further remedial activities.

4.4. SAFETY ASSESSMENT AND MODELLING

This section provides information and discussion on approaches for the safety assessment[15], including risk assessment[16] and environmental modelling of the potential radiological impacts on people and the environment posed by legacy trench sites.

Assessment and modelling will be iterative as additional information becomes available, for example, from site characterization and monitoring data or improvements in the interpretation of modelling results and model calibration.

Legacy trench sites are a unique subset of near surface disposal facilities owing to the characteristics outlined in Section 2.2. It is important that these features be adequately conceptualized and modelled to ensure appropriate assessment of the potential risks posed by these sites.

---

[15] The term 'safety assessment' is used by the IAEA to refer to all assessments performed as part of the safety case [57]. This will normally include risk assessment [25].

[16] Risk assessment is the assessment of the radiation risks and other risks associated with normal operation and possible accidents involving facilities and activities. This will normally include consequence assessment, together with some assessment of the probability of those consequences arising [25].

A notable feature of legacy trench sites is that they often contain a mixture of radioactive and chemically hazardous wastes. Therefore, the safety assessment of the legacy trench site would need to take into account both the radiological and chemical impacts using suitable methodology and modelling tools (e.g. see Refs [58–60]). The presence of other non-radioactive chemicals in the trenches may also impact the degradation of wastes and/or containers and the speciation and mobility of contaminants in the subsurface environment. For example, high ionic strength pore waters may suppress sorption of radionuclides that are retained in the soil matrix by ion exchange mechanisms [61]. In contrast, the presence of miscible organic substances can enhance radionuclide mobility in groundwater through the formation of non-retarded anionic chemical species and colloids (e.g. see Refs [62–64]). Such scenarios can become conceptually and computationally complex but nevertheless may become important pathways in the model.

4.4.1. Safety assessment

A safety assessment for legacy trench sites typically follows the same approaches used for near surface radioactive waste disposal facilities. The aim of the safety assessment is to evaluate the performance of the disposal system and quantify its potential radiological impact on human health and the environment [57]. The assessment usually considers the potential radiological impacts of a facility during its operation and after its closure. However, for a legacy trench site, the assessment is focused on evaluating the existing situation or to support a safety case to implement remedial activities (future situation). Where a predictive assessment is carried out, the output is compared with the respective regulatory requirements (e.g. reference levels). A risk assessment forms part of the safety assessment and is used to assess the risks posed to human health or the environment from the potential exposure to a contaminant. Recommendations on safety assessment are provided in the IAEA Safety Standard Series Nos SSG-29, Near Surface Disposal Facilities for Radioactive Waste [15], and SSG-23, The Safety Case and Safety Assessment for the Disposal of Radioactive Waste [57].

The IAEA developed a harmonized, iterative methodology for assessing the impact of near surface disposal facilities within the Improvement of Safety Assessment Methodologies for Near Surface Disposal Facilities project (ISAM) [65]. The follow-up project, Application of Safety Assessment Methodologies for Near Surface Radioactive Waste Disposal Facilities (ASAM), focused on evaluating the post-closure safety for waste disposal facilities and, where appropriate, the selection of corrective actions [66, 67]. The ISAM methodology (reproduced in Fig. 17) has gained widespread acceptance and

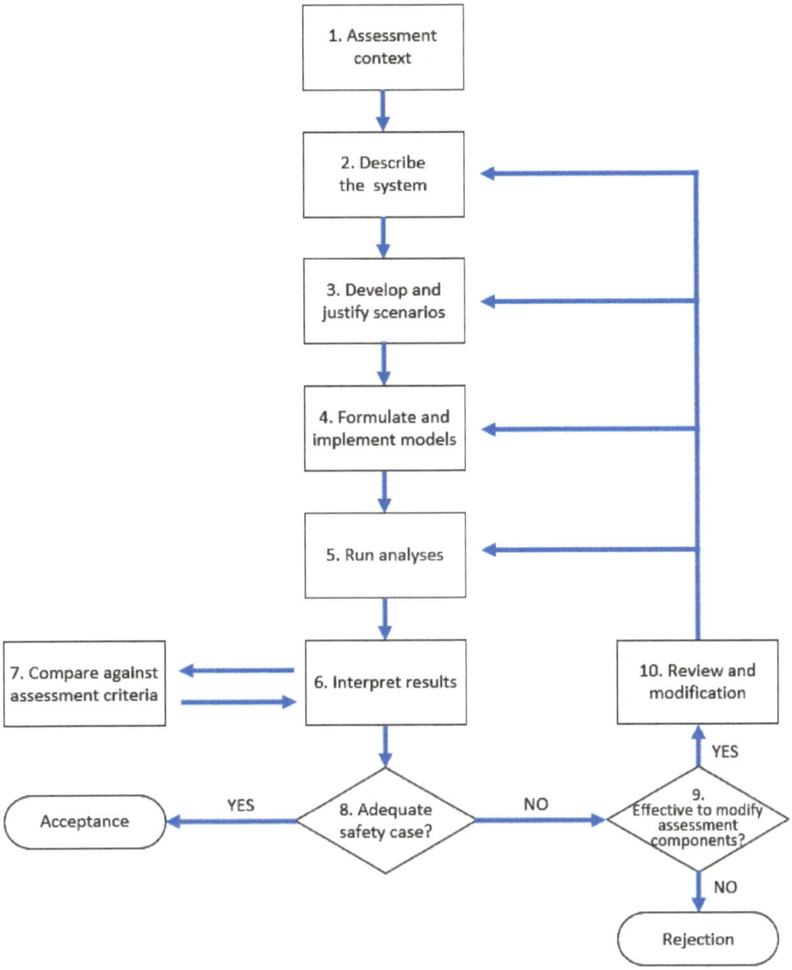

FIG. 17. The ISAM project methodology. Reproduced from Ref. [65].

has been proved to be a versatile tool for a range of applications, including the assessment of legacy trench sites (e.g. see Refs [12, 65, 68]).

References [65, 69] provide a description and test cases of assessment methodologies used for near surface disposal facilities, including a review of relevant source term models, contaminant release mechanisms, exposure pathways and radiological impact assessments. Many elements of these methodologies are directly applicable to the assessment of legacy trench sites. For a legacy trench site, these methodologies may need to be supplemented to include site specific factors and pathways and may require the use of non-standard scenarios, calculation procedures and/or models. These topics have been examined in the IAEA MODARIA[17] programmes. Of particular interest is the BIOPROTA project, which aims to address key uncertainties in the long term impacts to people and the environment as a result of waste management practices. In 2018, the BIOPROTA project published an update to the BIOMASS methodology; the methodology provides guidance on the consideration of the biosphere in post-closure safety assessments for solid radioactive waste disposal [70].

4.4.2. Modelling

Mathematical models are the mathematical interpretation of the conceptual model and can be used in several ways to support safety assessment and risk assessment and the development of a remediation solution. Models can be used as tools in the following applications:

- Quantitative description of the site condition (e.g. flow regime, contaminant source term, contaminant transport, geochemical conditions);
- Modelling contaminant transport (e.g. testing hypotheses on parameters and processes controlling the transport and fate of radionuclides from the contaminant source);
- Identifying gaps in site investigation or monitoring activities (e.g. the model predicts a plume into an area where there are currently no groundwater monitoring boreholes);
- Validation of hydrogeological parameters (e.g. the model contaminant concentrations are significantly different from the field measured contaminant concentrations);
- Planning future site investigation activities (e.g. using the model to identify which pathways and parameters have the greatest impact on the potential risks posed by the site);
- Evaluating the performance of the current facility (e.g. to ensure safety and regulatory compliance or to enable greater efficiencies);
- Enabling exemption and clearance of material with low levels of radioactivity;
- Evaluating the performance of proposed remedial options;
- Optimization of site monitoring systems.

[17] MODARIA stands for Modelling and Data for Radiological Impact Assessment. The MODARIA programme ran from 2012 to 2015, and the MODARIA II programme ran from 2016 to 2019 (see https://www-ns.iaea.org/projects/modaria/modaria2.asp?s=8&l=129). Work is currently ongoing to develop the scope of a future MODARIA programme.

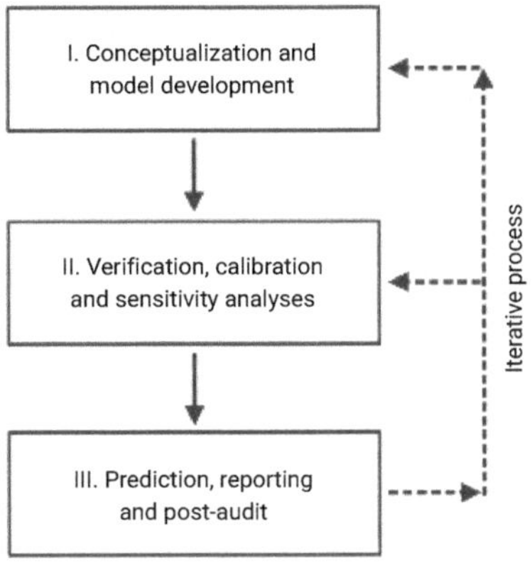

FIG. 18. Main stages of the model development and application process (based on Refs [71–74]).

Figure 18 outlines the main stages in the model development and application process. This methodology is valid for all the model applications listed above and is suitable for any type of contaminant transport models (e.g. atmospheric dispersion or surface water contaminant transport models).

The IAEA has provided support to Member States on environmental modelling through several programmes. The most recent were the EMRAS[18] and MODARIA programmes. The report from the EMRAS programme summarizes the outcome of testing models used in different contaminant transport scenarios [70, 75].

There is commercially available software, as well as free and open source software, that can be used to model potential radiation exposure and environmental impact from radiologically contaminated sites, including legacy trench sites. Examples of modelling platforms include GoldSim[19] and AMBER[20]. GoldSim was used to model contaminant transport as part of the safety assessment for the RADON type facility at Saakadze, Georgia [76]. GoldSim was also used to assess the groundwater, gas, coastal erosion and human intrusion pathways for the Safety Case at LLWR, UK [77].

---

[18] EMRAS stands for Environmental Modelling for Radiation Safety. The EMRAS programme ran from 2003 to 2007, and the EMRAS II programme ran from 2008 to 2011 (https://www-ns.iaea.org/projects/emras/emras2/default.asp?s=8&l=63).

[19] https://www.goldsim.com/Web/Products/GoldSim/Overview/

[20] https://www.quintessa.org/software/AMBER

Examples of radiological assessment tools include NORMALYSA (NORM And LegacY Site Assessment)[21], RESRAD (RESidual RADioactive)[22] and RCLEA (Radioactively Contaminated Land Exposure Assessment)[23]. RESRAD was used to calculate radiation dose rates to receptors for a range of medium and long term management options under consideration at the LFLS, Australia [78].

The groundwater pathway is a key exposure route when modelling disposal facilities and this is also often the case for legacy trench sites. Many resources outline the basic principle and approaches to modelling groundwater flow[24] (e.g. see Refs [79–84]) and pollutant fate and transport[25] (e.g. see Refs [85–87]). Examples of groundwater transport models include PAGAN (Performance Assessment Ground Water Analysis of low-level Nuclear waste) [88], MODFLOW (Modular Three-Dimensional Finite-Difference Groundwater Flow)[26] [89] and MT3D-MS (Modular 3-Dimensional Transport model – Multi Species) [90] and ConnectFlow[27] [91]. The PAGAN code was used to derive the upper bound estimates for strontium-90 transported from legacy trenches within the Chornobyl Exclusion Zone [92]. The VisualMODFLOW code, which is the combination of MODFLOW and MT3D, which is the predecessor to MT3D-MS, was used to model the groundwater pathway for the legacy trenches at the Ezeiza site in Argentina (see Appendix I, Section I.1). ConnectFlow was used to model groundwater flows as part of the Safety Case for LLWR, UK [77].

Assessment results, based on either measured or modelled data, will require evaluation in terms of risk and dose criteria, as relevant to the local regulatory regime. This is likely to involve a consideration of features, events and processes and of how these may change over time. Uncertainties will need to be addressed and sensitivity analysis is an important part of this (see Section 4.4.2.2).

4.4.2.1. Conceptualizing the site

For the purpose of conducting a safety assessment, the further development of the CSM and the creation of exposure scenarios is one of most critical and important phases of the assessment [65, 72].

---

[21] http://project.facilia.se/normalysa/index.html

[22] http://resrad.evs.anl.gov/

[23] https://www.gov.uk/government/publications/rclea-software-application

[24] Groundwater flow models calculate the direction and rate of groundwater movement through an aquifer or confining units in the subsurface.

[25] Fate and transport models calculate the chemical fate of contaminants as they move through the groundwater.

[26] https://www.usgs.gov/mission-areas/water-resources/science/modflow-and-related-programs?qt-science_center_objects=0#qt-science_center_objects

[27] https://www.nammu.co.uk/

Building on the source–pathway–receptor methodology (see Section 4), the CSM can be used to describe the site in terms of features, events and processes[28]. The relationship between the features, events and processes determines the contaminant transport and fate and their impact on human health and the environment[29]. Thus, the model consolidates the current understanding of the key processes influencing the legacy trench site (now and in the future) and describes the possible response and evolution of the trenches and related impacts. This may, for instance, include consideration of climate change, degradation of the waste form or types of intrusion or physical disturbance.

Typically, the CSM is subdivided into the near field[30], far field[31] and biosphere subsystems, and then the features, events and processes are modelled within and between the subsystems. Two methods that are frequently used to map the interactions and pathways of features, events and processes are the interaction matrix approach and the influence diagram approach [65]. Both approaches provide a formalized systematic process enabling visualization of the interactions and pathways.

It is important to recognize that the CSM for a legacy trench site will have differences compared with one prepared for a modern operational disposal facility. For a legacy trench site, key pathways may include groundwater bath-tubbing, biomigration and direct exposure resulting from the poor construction of the facility. At a modern facility, the key pathway may be groundwater migration. An example of an important feature for a legacy trench could be the absence of a trench cover; a process could be the degradation of the trench cover; and an event could be inadequate controls to limit site access, leading to the exposure of a defined reference individual to contaminated material in the legacy trench. This feature, event and process combination could also lead to a higher probability of an intrusion scenario. Another legacy trench site process may be the presence of decaying organic material in the trenches. This could affect the geochemical conditions (e.g. redox, solution chemistry, pH, organic

---

[28] Features, events and processes are often referred to as 'FEPs'.

[29] This includes physical features, events and processes that could directly or indirectly influence the release and transport of radionuclides from the repository or subsequent radiation exposures to humans, as well as other factors [65].

[30] The IAEA Safety and Security Glossary [25] defines the near field as "The excavated area of a *disposal facility* near or in contact with the *waste packages*, including filling or sealing materials, and those parts of the host medium/rock whose characteristics have been or could be altered by the disposal facility or its contents."

[31] The IAEA Safety and Security Glossary [25] states that far field is "The geosphere outside a *disposal facility*, comprising the surrounding geological strata, at a distance from the *disposal facility* such that, for modelling purposes, the *disposal facility* may be considered a single entity and the effects of individual *waste packages* are not distinguished."

content) and have an effect on the release rate and mobility of contaminants to groundwater. Systematic descriptions of radionuclide release, transport and exposure mechanisms and factors related to the above pathways can be found in Refs [65, 69]. Appendix II provides further detail on key exposure pathways and scenarios for legacy trench sites.

The level of complexity of the CSM needs to be consistent with the assessment context, including the assessment objectives, the availability and the quality of site data, as well as access to modelling expertise and resources. In conditions of limited site data and at the preliminary (screening) stages of assessment, use of relatively simple conceptual and mathematical models (analytical, semi-analytical, compartmental models) is often justified. Such models are most useful for getting upper bound (worst case) estimates of radiological impacts. However, simplifying assumptions in models need to be used with caution, as assumptions that are conservative for one pathway (or scenario) may be non-conservative for another.

The development of the CSM is iterative and adaptive and needs to respond to the increase in information and experience over time. For example, the early assessment of the Chornobyl Red Forest radioactive waste burials assumed that groundwater migration was the primary pathway and used a simple source term model, which assumed that all the radioactive inventory in the trenches was in a mobile form. As a better understanding of the site was developed from the site characterization and monitoring studies (2000–2009), a complex source term model was devised, which accounted for radionuclide association and release from the fuel 'hot particles' in waste material. In addition, the geochemical evolution of the site due to degradation of organic matter buried inside the trenches was recognized as an important factor causing attenuation of strontium-90 migration [61, 93]. The most recent safety assessment has identified that external exposure to the legacy trenches and biomigration of radionuclides are also important pathways. These results have subsequently been addressed in the planned remedial measures for the site [94].

A discussion of the general approaches and key issues for developing CSMs for safety assessment can be found in Refs [65, 73, 74, 79].

4.4.2.2. Uncertainties

As discussed in Section 2.2, there are many uncertainties that need to be addressed in the assessment of a legacy trench. Uncertainty is recognized as a key issue in the development and application of safety assessments and environmental modelling for radioactive waste repositories (e.g. see Refs [57, 74]). The ISAM project methodology notes that "The identification of sources of uncertainties as well as the types of uncertainties are necessary in order for the analyst to find the best way to quantify and consequently improve the degree of confidence he or she has in the safety analysis" [65], and distinguishes the following types of uncertainty:

- — Scenario uncertainty: uncertainty related to the long term future behaviour of the disposal facility. It includes human use of the land, geo-hydrological processes, intrusion and other long term processes. To address this uncertainty, it is necessary to analyse a comprehensive range of scenarios that cover the likely evolution of the site.
- — Model uncertainty: uncertainty in how physical and chemical processes are represented in a mathematical model. In order to address this type of uncertainty, several alternative conceptual and/or mathematical models can be explored in parallel.
- — Data and parameter uncertainty: uncertainty that stems from the potential lack of reliable site specific parameters that are needed to run a mathematical model. This type of uncertainty can be addressed by using bounding estimates or probabilistic modelling framework.
- — Subjective uncertainty: uncertainty arising from the need to rely on expert judgement owing to lack of data, lack of knowledge concerning future conditions and parameter values (and distributions), or any aspects of the system under study that are not well understood by current science.

Conservative analysis

In this approach, conservative (reasonably pessimistic) parameter values are chosen to ensure that the modelled impact resulting from the source term (e.g. the legacy trenches) is not underpredicted. Therefore, modelling provides an upper bound estimate of the likely dose or risk from the legacy trench site.

There may be cases where a conservative analysis predicts impact from the source that is clearly unrealistic. In these instances, an analysis approach that provides a greater level of detail is needed. Caution is needed, because an unrealistically conservative impact assessment of a site may result in an over-predicted impact, which may lead to unnecessary remedial activities with associated health and safety, environmental and financial impacts.

Sensitivity analysis

Sensitivity analysis is carried out by systematically altering model parameters within the plausible range and assessing the effect on the model output. In this way, the most sensitive and hence key model parameters can be identified. Sensitivity analysis also allows the prediction of the range of potential impacts from the site, rather than a single value. In other cases, insensitive parameters can be flagged which may save time in evaluating unnecessary model components. This procedure can provide feedback to model the parameterization process, site characterizations and monitoring programme, and the development of remedial options [65, 74]. The two primary methods of sensitivity analysis are as follows:

- — Bayesian updating for model calibration. In this approach, the model parameters that do not allow adequate reproduction of historical data are assigned reduced or even zero probabilities. Thus, the uncertainty of the a priori model parameters is revised during model calibration, resulting in a posteriori model parameters with reduced uncertainty [74, 84].
- — Probabilistic Monte Carlo simulation. In this approach, the model parameters are treated as uncertain variables, which are characterized by relevant statistical probability distributions. The Monte Carlo method is used to propagate the uncertainty in the input parameters to generate the model output. The Monte Carlo method is used to generate many sets of random input parameters to enable the model to generate a range of outputs derived from the calculated combination of the many inputs. The statistical characteristics of model outputs can then be analysed (e.g. mean, median values and/or confidence intervals).

Additional discussion of methods used in modelling to address uncertainty can be found in Refs [65, 74, 80].

5. DECISION MAKING AND THE END STATE

Legacy trenches do not typically conform to current standards for disposal facilities, but that does not automatically mean that the material inside the trenches has to be recovered or that major interventions are necessary. Excavating and removing large volumes of radioactively and chemically contaminated material from legacy trenches for disposal elsewhere is likely to involve logistical and technical challenges, including risks to workers and potential environmental impact.

Decision making on the need for corrective action to achieve an acceptable state for a site will generally focus on one or more of the following questions:

- — Is the existing condition of the trench unsafe or unsatisfactory, or does the trench have the potential to give rise to an unsafe or unsatisfactory condition in the future? (See Section 4.4.)

— What are the options for intervention? (See Section 6.)
— Is intervention needed to respond to societal concerns or demands? (See Section 7.)
— Does the applicable regulatory body require action? (See Section 2.5.)
— What is the intended future use and end state of the site?

It is recognized that the challenge of defining the site end state is not just a technical challenge; the process requires the consideration of environmental, social and economic factors and the inclusion of interested parties in the decision making process.

This section outlines how decision making for the management of legacy trenches can be integrated with the determination of an agreed site end state, with the aim of optimizing the remediation approach to achieve an acceptable condition. It is based on the strategy of employing early, risk based decision making to ensure that remediation is proportionate to the level of risk and to ensure that its objectives are met in ways that minimize waste generation and worker dose.

5.1. END STATE OBJECTIVES

The term 'end state' is defined as a "predetermined criterion defining the point at which a specific task or *process* is to be considered completed" [25]. For the purpose of this publication, this definition is clarified as the site specific condition to be achieved by decommissioning and/or remediation that reflects the intended future use of the site and is appropriately protective of people and the environment. This definition is in line with the IAEA guidance concerning the determination of environmental remediation end states [4], and that developed by the UK Nuclear Decommissioning Authority (NDA), who has spent recent years updating the site end state strategy for its nuclear sites that include legacy trenches. The NDA states that the "*site end state* describes the condition to which the site (land, structures and infrastructure) will be taken at the end of the decommissioning process" [95].

The first task for any site end state project is to ensure that all interested parties, including the project team, understand what the end state means in the context of their site. In some instances, the end state could be linked with the release of the site from regulatory controls, meeting a specific regulatory milestone, the transfer of liabilities between organizations or the completion of monitored natural attenuation.

The site end state is broader than just the future use of the site. It includes the following:

— The physical state of the site (e.g. the inventory of waste and/or land contamination remaining at the site) and the visual appearance and associated infrastructure.
— Where necessary, the controls[32] to protect people and the environment from any residual hazards.
— When and how the site can be used. The use may be restricted or unrestricted; restrictions may be for a defined period or may be permanent.

For example, a site may be used for industrial purposes whether land contamination is present or not. In this instance, the site end state would differ because controls would be needed if the land contamination remained at the site to ensure that the site remains safe for the intended use.

Traditionally, the next use of the site has been a primary factor in defining the site end state; however, for legacy sites, other factors may also be prominent in the decision making process. These factors include the environmental setting and the distance to urban areas; economic opportunities for reuse; social context, including local acceptance; availability of funding; availability of waste routes; and political context. The preferred site end state option will be selected through a process of optimization that takes into account the range of factors that are relevant for the site. However, it is important to acknowledge that retaining a legacy trench at a site is likely to have an impact on the potential future uses for the site. This is also true for any site with a waste disposal facility. At the strategic level, the options of the site end state are the following:

— Managed in situ in perpetuity;
— Closed waste management facility;
— Legacy trenches are removed, and associated ground and groundwater contamination is remediated.

[32] Physical controls; this includes engineered controls (e.g. capping layers, subsurface containment walls, fences) and non-engineered controls (e.g. depth to contamination, natural geological feature).

There are four technical approaches to address a legacy trench site to meet the preferred end state; these can be summarized as follows:

— In situ management without intervention — where underpinned by a safety case and acceptable on the principles of best available technique (BAT) and/or ALARA.
— In situ management with intervention — where necessitated by the principles of BAT or ALARA to break source–pathway–receptor linkages; for example, capping based on human intrusion risk, engineered in-ground barriers based on groundwater risk, or treatment to encapsulate by grouting, vitrification or injection (chemical and/or biological).
— Ex situ excavation for on-site disposal; material is removed, conditioned and disposed of in an engineered waste disposal facility on-site. The facility may be a newly constructed facility or a repurposed existing facility such as a former basement.
— Ex situ excavation for off-site disposal[33].

Depending on the selected site end state, there are two approaches to ex situ excavation, as follows:

(a) Selective removal of contamination 'hot spots' based on human intrusion risk, presence of specific contaminants (e.g. long lived radionuclides, mercury) or mobility or risk to groundwater;
(b) Comprehensive excavation.

Where a legacy trench site is part of a larger nuclear or industrial site, it will be necessary to consider the end state for the trench area in the context of the whole site end state. It may be appropriate to consider different strategic options for different areas of the site to facilitate the next use. The site end state will require integration of the end state for facilities and contaminated land areas to form the overall site end state. To achieve this balance, it might not be possible to select the preferred end state for every facility or contaminated land area but instead to reach an optimized solution that balances the overall preferred site end state.

Where national, local or site policies or strategies are not mature, it may be difficult to define the site end state. For example, a national waste policy might not exist, or on a large site there may still be uncertainty regarding ongoing use of the site for nuclear purposes. In such cases, it may be preferable to select an interim state rather than an end state. An interim state can be defined as "typically

---

[33] Off-site disposal is where the waste is removed from the generating site and disposed of at a disposal site located elsewhere.

a stable state that marks a stepped reduction in risk or hazard, and may be associated with a reduction in regulatory controls" [95]. Interim states are natural milestones and decision points and typically indicate a reduction in risk or hazard. The use of interim states is likely to affect the implementation approach, affordability, lifetime cost and timescale of achieving the site end state. It may be necessary to optimize the selection of interim states to ensure that they bring an overall benefit to achieving the end state.

An example of the use of an interim state is the current approach used to manage the Windscale Trenches at the Sellafield site [50]. It is expected that decommissioning, remediation and waste management activities will be ongoing at the Sellafield site until 2120 [96]. Therefore, improvements to further limit the potential ingress of rainwater were made to the Windscale Trench site to enable an interim use of the capped area itself and ensure ongoing protection of people and the environment. Other reasons that may require the selection of an interim state include the lack of a suitable waste route or insufficient resources to complete the work.

5.2. ASSESSMENT FRAMEWORK FOR DECISION MAKING

There is currently limited published guidance specifically on decision making to support identification of the site end state, although there is growing experience in Member States. The IAEA has recently published guidance on the determination of environmental end states [4], and guidance published by the NDA [97] and the US DOE [98] is also useful for end state decision making.

Figure 19 presents the end state decision making process in the context of the remediation process (see Section 2.4). This diagram is aligned with Ref. [4] and GSG-15 [2].

An assessment approach needs to be established that is proportionate but sufficiently structured to provide confidence to interested parties that the decision is robust and underpinned. An essential part of the process is the definition of assumptions and constraints; all assumptions and constraints need to be clearly documented as part of the decision making process.

Assumptions enable the management of uncertainties in the decision making process. Assumptions are a way of fixing a variable to allow an option to be defined and assessed. Typical options in the context of legacy trench site decision making include the following:

- — Regulatory requirements will not change over time;
- — Key skills and resources will be available as necessary;
- — A viable waste route is available if material needs to be removed from the site.

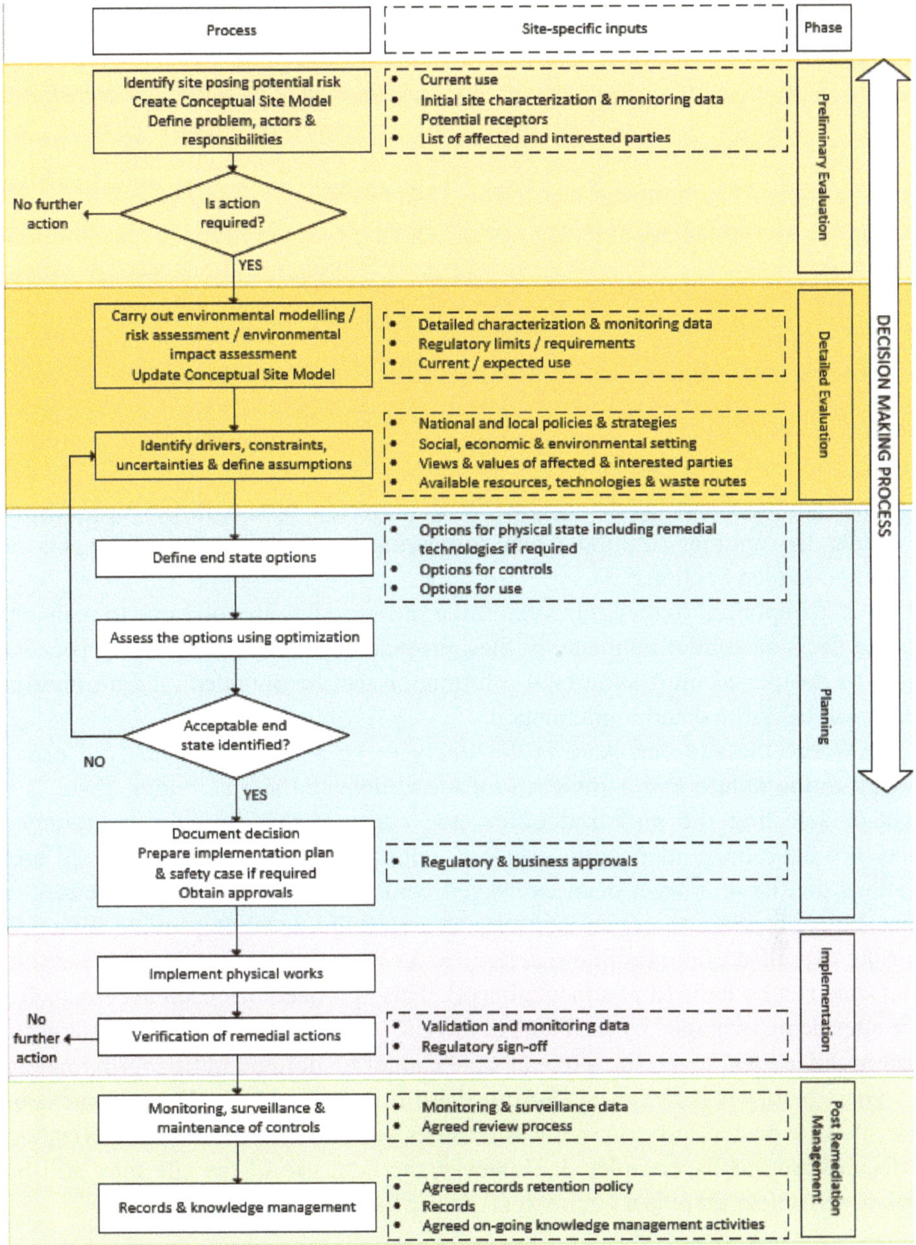

FIG. 19. Decision making process in the context of the remediation process.

A constraint restricts what can be achieved. Constraints do not limit the range of options, they only restrict the implementation of the option. Constraints can be 'hard' or 'soft', and it may be possible to change or remove a constraint. A soft constraint is something that could be relatively easy to change. For example, there is not enough space to apply a certain remedial technology — this constraint could be removed if additional land could be made available. A hard constraint is something that may be difficult, or impossible, to change. For example, a regulatory requirement. However, there may be instances where the value of removing even a hard constraint can provide long term and/or wide reaching benefits. This is the case in the UK, where outdated legislation is being changed to support a pragmatic approach to regulating nuclear sites at the latter stages of decommissioning and remediation (see Refs [99, 100]).

Identifying the preferred option can occur when uncertainties have been reduced to a level where there is confidence (from interested parties) that the optimal solution can be determined. The preferred option is the option that provides the optimal solution when comparing the benefits and detriments of each option (see Section 5.3).

It is important to recognize that if the uncertainties are too large to manage, then a decision cannot be made. In these instances, the decision making process may be postponed until additional information can be obtained, and an interim state will be defined and implemented.

Where the site end state is not likely to be achieved for many decades owing to the nature and complexity of the anticipated site activities, there is a risk of selecting the preferred option too early. If this occurs, uncertainties may not have been adequately resolved, optimization may be constrained and options that have not yet been envisaged could be ruled out. In these cases, it may be appropriate to set an end state at a strategic level to provide direction for the overall decommissioning activities. As the uncertainties are reduced, the end state can be defined at a tactical level. This approach has been necessary for the sites being managed by the NDA in the UK [95]. The UK experience suggests that at the tactical level, the site end state cannot be defined until approximately 15 years before it is achieved. Within these timescales, the political context of the site is expected to be stable, waste routes are identified and characterization activities are mostly completed. However, the next use of the site may still be unknown, unless there is a commercial driver for reusing the site.

5.3. OPTIONS ANALYSIS

Careful consideration needs to be given to identifying the factors — sometimes termed attributes — used to assess end state options.

Although there may be a 'standard' list, these factors need to be specific for the decision being made. In line with the principle of optimization and sustainability, factors need to cover social, environmental and economic aspects, as well as technical requirements.

Examples of social factors can include the following:

- — Supporting the reuse of the site for the local community;
- — Creating the largest number of jobs for the local community during the remedial works;
- — Removing blight[34] associated with the site.

Examples of environmental factors include the following:

- — Nuisance caused by transport of materials to and from the site;
- — Improvement in the diversity of flora and fauna;
- — Preventing migration of contaminants to groundwater resources.

Examples of economic factors include the following:

- — Delivering lowest lifetime costs;
- — Presenting the lowest financial risk by using proven technologies.

Examples of technical factors used to assess the options include the following:

- — Meet the requirement to break a pollutant linkage (see Section 4);
- — Can be implemented with certain equipment;
- — Can be implemented in a preferred timescale.

It is worth noting that some factors could be classified into more than one category, depending on how they are defined. Different categories could also be used; for example, factors with organizational impact (i.e. internal impact) such as resource needs, and external impacts, such as environmental benefits. The specific terminology and grouping of factors will be influenced by the

---

[34] In this context, 'blight' refers to the perception of contaminated land and whether the presence of the site impacts on whether people wish to live near the site. Concerns may be associated with the visual appearance of the site and the real or perceived health and safety risks. A proxy for measuring this impact is to look at the rate of property sales and the property prices in the area.

project team and the interested parties and needs to be agreed before undertaking the assessment.

Sources such as Refs [7, 97, 101, 102] provide a starting place for the development of a list of factors. Factors must be discrete to avoid double counting. It may also be appropriate to exclude factors that do not discriminate between options. However, it may also be necessary to include these factors if they are required to support the specific needs of interested parties.

The options assessment process may be phased to reduce the number of options being considered and, where necessary, support the development of options or suboptions. The initial screen assessment can be used to do the following:

— Clarify uncertainties and associated assumptions;
— Remove options that do not meet the constraints;
— Identify options that require additional information or further analysis;
— Develop the approach for future detailed options assessment.

The factors may be scored qualitatively or quantitatively on the basis of the type of factor, the available information and the assessment methodology being used. There are several methodologies or tools available for decision making, including direct evaluation, the linear additive method, multi-criteria decision analysis[35] and cost–benefit analysis[36]. References [98, 103–109] provide a wider discussion about the selection of decision making methods and tools and include summaries of commonly used methods and tools. The method selected needs to be proportionate to the complexity of the decision.

5.4. ITERATIVE APPROACH

Although Fig. 19 illustrates a predominately linear process, the process of defining the site end state will be iterative and adaptive to accommodate new information and changes in the context of the decision and/or assumptions. This iterative process is likely to be more pronounced when the site is complex and/or decommissioning and remediation activities will take many years. As discussed above, the use of interim states can help to manage the iterative nature of the process, especially when there are large uncertainties.

In the early stages of determining the site end state, the work will be focused on studies to collect missing information. As the work progresses, the focus of the

[35] Often referred to as MCDA.

[36] Often referred to as CBA.

work may lead to the development of a greater understanding of the contaminant behaviour and the development of additional features, events and processes. In some sites, research and development may be needed to further underpin the site end state options. The process needs to be documented as a means of establishing a decision record to ensure due process and demonstrate transparency to support engagement with interested parties. This record can also be preserved to assist in future corrective action evaluations.

The term 'adaptive site management'[37] is used in North America to describe the iterative and adaptive approach to managing complex or difficult to remediate sites. The USA ITRC (Interstate Technology and Regulatory Council) provides guidance on the remediation management of complex sites and has documented a process for adaptive site management that provides an example of a structure to remediate complex sites [111, 112].

5.5. PERFORMANCE OF CORRECTIVE MEASURES

Following the completion of physical works to achieve the site end state, monitoring may be necessary to evaluate the performance of remedial measures. Indicators are used as a method to confirm that the corrective actions have achieved their intended purpose. It is therefore important to create a site condition baseline prior to implementing the corrective actions to provide a basis to assess the effectiveness of the remedial measures. At a legacy trench site, performance indicators may include a reduction in contaminant concentration at a compliance point, a reduction in leachate production and the stability of trench cover contours.

In some cases, it can be difficult to measure the effectiveness of corrective action over a short time frame. For example, experience has shown that monitoring and safety assessments are an alternative approach to demonstrate the reduction of long lived radionuclides compared to validation measurements following the completion of the remedial activities. It is important to discuss the performance measures with interested parties to ensure that there is agreement over which indicators will be used and the type of results that are expected over certain timescales.

---

[37] The USA ITRC uses the following definition for the term 'adaptive site management' as defined by the US National Research Council: "an approach to resource management in which polices are implemented with the express recognition that the response of the system is uncertain, but with the intent that this response will be monitored, interpreted, and used to adjust programs in an iterative manner, leading to ongoing improvements in knowledge and performance" [110].

6. REMEDIATION

Remediation is defined as "Any measures that may be carried out to reduce the *radiation exposure* due to existing *contamination* of land areas through actions applied to the *contamination* itself (the *source*) or to the *exposure pathways* to humans" [25]. Any measure to break a pathway can be considered as remediation; for example, restricting access to the site or placing an impermeable cap over the legacy trenches to limit rainwater infiltration. However, the term 'management' rather than 'remediation' can be used to indicate that the site will continue to be actively monitored and that the site end state or site closure has not been achieved.

Remediation and/or management approaches at legacy trench sites will need to take into account both the legacy trenches and any impacted ground and water bodies. If there is currently little or no contaminant impact to the ground and groundwater surrounding the trenches, remedial activities will be focused on the legacy trenches themselves. In this instance, either waste could be retrieved (see Refs [3, 13] and Section 6.1) or remedial activities could improve the legacy trenches to enable the waste to remain in situ (see Ref. [12] and Section 7.2). Where the surrounding environment has been impacted by the migration of contaminants, remedial activities will also need to address soil and groundwater contamination (see Refs [72, 113–118] and Section 6.3).

It is important to consider all the available applicable technologies before starting remediation. Applying an inappropriate remediation approach could have a detrimental effect on the environment and misuse funds, especially if additional works are subsequently needed to rectify the situation. Adopting a phased approach to any remediation programme, including the consideration of smaller scale pilot studies, can also enhance the chances of success.

Table 4 provides a summary of where the different technologies have been applied internationally to remediate legacy trench sites. The following sections provide a discussion of the remedial approaches and their application.

6.1. WASTE RETRIEVAL

At several legacy trench sites, the trench content has been retrieved and conditioned for storage and/or disposal. A key driver for this approach could be the need to reuse the site. For example, at the Southern Storage Site in Harwell, UK, all the wastes and contaminant impacted soils were removed and the site has subsequently been reused for residential housing [28]. Waste retrieval has also been adopted at several sites in the USA. At the Oak Ridge Reservation, there

TABLE 4. SUMMARY OF TECHNOLOGIES USED TO REMEDIATE SELECTED LEGACY TRENCH SITES

Approach	Technology type	Site
Waste retrieval	Not applicable	Harwell (UK), Rocky Flats (USA), Los Alamos (USA), Oak Ridge (USA), Idaho National Laboratory (USA), Ezeiza (Argentina)
In situ management	Engineered barriers	Maralinga (Australia), Centre de la Manche (France), Sellafield (UK)[a], LLWR (UK)[a], Oak Ridge (USA), Maxey Flats (USA), West Valley (USA), Beatty (USA), Sheffield (USA), Ezeiza (Argentina)
	In situ stabilization	Maralinga (Australia), Oak Ridge (USA)
	In situ treatment	Oak Ridge (USA), Rocky Flats (USA), Idaho National Laboratory (USA)
	Monitored natural attenuation	Red Forest (Ukraine)
	Hydraulic control	—

[a] This is viewed as an interim state prior to the application of the final remediation option.

are many legacy trench sites; differing remedial approaches have been applied at different sites including excavation. For example, at the K-1070-B burial ground site, six trenches and the surrounding soil were excavated to approximately 3 m below ground level to protect groundwater [119]. Retrieval works were completed in 2019 to remove waste from the Pisky-1 decontamination waste storage facility in Ukraine [120]. The site is located outside the Chornobyl Exclusion Zone and was established for the disposal of wastes generated from the cleanup of the surrounding territory. Pisky-1 was selected as a pilot project, following the safety assessment of the eleven decontamination waste storage facilities [9, 121].

It is very likely that exhumed waste will require pretreatment and/or treatment and conditioning prior to storage and/or disposal. Reference [13] provides guidance on waste retrieval approaches and Ref. [3] provides guidance on the technical considerations for waste processing activities.

6.2. IN SITU MANAGEMENT OF WASTE

Decision making underpinned by a safety case may support either the ongoing management or closure of the site with corrective actions. In situ corrective actions can be organized into four categories: engineered barriers; in situ waste stabilization; in situ waste treatment; and water collection and extraction. Further information on these approaches can also be found in Ref. [12].

6.2.1. Engineered barriers

Surface caps are a common approach for the containment of contaminated materials and wastes [122]. Caps are placed over the trenches to isolate them from natural erosive processes and rainwater infiltration. Caps can also help to control any gas or vapour emissions and limit intrusion from burrowing animals. The design of the cap will vary depending on the waste inventory and the local climatic and hydrogeological conditions. Caps may primarily consist of clay, local soil, concrete, geomembrane or they may be multilayered in design. Different types of surface cap have been applied at many sites, including Sellafield (UK), Maxey Flats (USA), Maralinga (Australia), Chalk River (Canada) and Centre de la Manche (France). At the Maxey Flats site, the trench materials were firstly allowed to subside naturally to a stable condition and then a permanent engineered cap was placed over the area [123].

Impermeable cut-off walls are subsurface vertical barriers designed to direct or impede groundwater flow. They may be used to direct contaminated water from the legacy trenches or to direct up-gradient groundwater away from the legacy trench site. Although they do not prevent the vertical migration of contaminants, they can inhibit lateral migration from the trench area. Their implementation is highly dependent on the physical properties of the geology and the specific depth required to control the contaminant migration. Such cut-off walls may include sheet piling walls, biological barriers, cement based grout curtains, soil freezing and bentonite slurry walls. At the West Valley site in the USA, a 250 m long groundwater barrier wall was constructed along two sides of the waste trench area. This barrier wall involved the excavation of a trench into undisturbed soil, which was then backfilled with a combination of native soils combined with bentonite to form a slurry, thus forming a low permeability barrier [124]. At LLWR, UK, in 1984 tritium was discovered in groundwater samples to the east of the railway drain on the east of the site. This is indicative of groundwater flow through the legacy trenches to the east. To eliminate, or reduce, such flows, a cement bentonite slurry cut-off wall on the north and east of the trenches was installed in 1989 and extended alongside the trench in 1995. The performance of the cut-off wall is monitored through both surface water and groundwater monitoring. The cut-off wall is expected to be extended around the disposal area and will act as a barrier for lateral migration both into and out of the disposal area [125].

Below ground barriers constructed of impermeable material can be emplaced beneath the waste form. The purpose of such barriers is to prevent the vertical migration of contaminants into the groundwater. The installation of such barriers will be greatly influenced by the physical characteristics of the soil. Examples of below ground barriers include liners (a technology commonly utilized at conventional waste landfill sites), grout injection and polymeric barriers. However, the installation of below ground barriers is difficult and may involve excavation of the waste, which may pose a risk to workers.

6.2.2. In situ stabilization

Stabilization technologies are aimed at reducing the mobility of contaminants and preventing their migration from the source area [12]. In situ stabilization can be used to stabilize the waste if the waste containers have degraded and no longer provide containment. These technologies are often split into two subareas, namely, in situ encapsulation and compaction.

In situ encapsulation involves the injection of grout or a polymer to immobilize any contaminants within the waste form. This approach, if effectively implemented, can create a monolithic structure that provides stability inside the trenches and limits the mobility of water through the wastes. However, it can be difficult to implement this technology to ensure that the encapsulant fills all the voids in the waste trenches. This technology was applied to legacy Trench 5 and Trench 7 in the Melton Valley at the USA Oak Ridge Reservation [126–128]. The original remedial approach for this area was in situ vitrification. However, it was later determined that in situ grouting with a cement based substance would be more effective and less expensive.

In situ compaction, on the other hand, can be used to reduce the volume of solid and semi-solid materials such as sludges, soils and other bulk materials. By reducing the voidage in the waste, the potential for settlement and potential damage to capping layers is reduced. This, in turn, improves the longevity of the cap and the subsequent risk posed by water penetrating into the waste. At LLWR, UK, compaction of the waste was partially achieved during placement, with 800 000 m^3 of the waste sent to the site occupying 500 000 m^3 of space in trenches. Construction of the final cap is expected to cause additional compaction of the waste, and by surcharging[38] ahead of construction, it is expected that the voidage will be minimized.

Solidification of semi-solid materials in situ can be achieved by slowly increasing the weight of material placed above the wastes. This will drive the water out of the waste, allowing settlement and/or compaction. Alternatively, cementitious material can be added through soil mixing or injection to solidify the waste and reduce settlement.

Compaction and solidification also reduce the permeability of the material, limiting the amount of infiltration and groundwater movement.

6.2.3. In situ treatment

Typically, in situ treatment technologies are applicable to the treatment of contaminated ground and groundwater, although some techniques have been applied to treat legacy waste trenches. In situ treatment technologies allow contaminated materials to be treated without extraction, transport or future redisposal. Their aim is to reduce the hazard posed by a legacy waste trench by immobilizing, destroying or removing contaminants. The success of in situ waste treatment technologies and techniques is difficult to verify owing to their limited application.

Physical and chemical treatment technologies can be used to chemically convert or separate contaminants in the trenches and/or the impacted ground surrounding the trenches. Flushing, for example, can extract contaminants from legacy trenches by injecting an extraction fluid, which is subsequently recovered. Chemical oxidation or reduction is undertaken by injecting chemical agents into the contaminated area to immobilize or destroy the contaminants. The objective is to transform any radionuclides into a less soluble form and to destroy any associated organic contaminants. At the Idaho National Laboratory (USA), unsaturated zone vacuum extraction and treatment were used to remove organic contaminants from the Radioactive Waste Management Complex trenches [129–131].

Thermal treatment relies on the application of high temperatures to reduce the mobility of contaminants. With in situ vitrification, high temperatures can be utilized to melt contaminated soil and sludges. A vitrified glass matrix is formed upon cooling, thus immobilizing any contamination. This method can be applied to waste trenches if the wastes themselves are predominantly composed of soil. Like physical and chemical treatment technologies, it would not work for bulk wastes. At the Maralinga weapons testing site in Australia, a series of waste

[38] Surcharging consists of applying load on the ground surface to accelerate consolidation of the materials in the ground.

pits were treated through in situ vitrification. However, because of an explosion during the vitrification process, this treatment was halted, and it was decided to exhume the wastes and place them in capped trenches [132].

6.2.4. Hydraulic control

Hydraulic control measures can be utilized to mitigate the spread of contamination by changing the localized hydraulic gradient. When contamination is deep, the impacted groundwater can be withdrawn through extraction wells or through near surface drains and ditches if the groundwater intercepts these features. This approach is more likely to be successful when applied to the more accessible surface drains and ditches. When the application is for deeper groundwater, there is a greater necessity to have a robust understanding of the local hydraulic conditions. This understanding will ensure that any extraction wells are suitably positioned, and that the most appropriate pumping rates are chosen. Impermeable cut-off walls can also provide a certain level of hydraulic control.

6.3. REMEDIATION OF GROUND AND GROUNDWATER

If the end state for the site involves the removal of the legacy trenches, any underlying or adjacent contaminated soil or groundwater may also need remediation. However, if the legacy trenches will remain in situ, it may only be necessary to demonstrate that any residual adjacent contamination, including potential future releases, is within regulatory limits and fulfils the requirements of the safety case. The remediation approach needs to be commensurate with the desired site end state and end use.

Several of the technologies described in Section 6.2 can be used to support the remediation of contaminated ground and groundwater. For example, an impermeable cut-off wall can be used to funnel contaminated groundwater either to an extraction point to allow ex situ treatment or towards a permeable reactive barrier for in situ treatment. In an ex situ approach, the groundwater is extracted and passed through a treatment plant, which is specifically designed to remove the contaminant(s) of concern. Typically, for dissolved metal contaminants such as cadmium, mercury or strontium-90, an ion exchange resin is used to remove and concentrate the contaminant. The resin is then disposed of as solid waste. At the Chalk River Laboratories site in Canada, an in situ permeable reactive barrier comprising clinoptilolite is being used to attenuate a strontium-90 groundwater plume [133]. Impermeable below ground barriers have been installed to funnel the groundwater through a permeable reactive barrier. The strontium-90 is bound to the reactive material, thereby reducing the concentration and migration of strontium-90 in the local groundwater.

Many sources describe the various approaches for remediating contaminated ground and groundwater, including those published by the IAEA and the Nuclear Energy Agency of the Organization for Economic Cooperation and Development (e.g. Refs [50, 72, 114–118, 134]) and on-line resources (e.g. Refs [135–137]). Table 5 provides a brief summary of remedial approaches and associated technologies used for radioactively contaminated ground and groundwater. Note that remedial technologies can be grouped in many ways; Table 5 combines the approaches used in Refs [114, 115]. Table 5 also records where the technology is applicable to non-radioactive contaminants.

TABLE 5. SUMMARY OF THE COMMON REMEDIAL APPROACHES

Approach	Technology	Application	Description and application
In situ partial or complete removal	Electrochemical remediation [138]	Soil; heavy metals, including radionuclides	A technique where a low direct current or potential gradient is applied between electrodes to induce the transport, transfer and/or transformation of contaminants. Movement will be towards an electrode, which can be located in a well to allow collection of the contaminants. The technique can treat both unsaturated and saturated soils. This is still a novel technique. It could also be used as an ex situ approach.
	Phytoremediation [139, 140]	Soil and groundwater; heavy metals, including radionuclides and hydrocarbons (e.g. oils, pesticides)	A treatment that uses plants to remove or stabilize contaminants. The process is most effective when the concentration of contaminants is low because high contaminant concentrations can inhibit growth of the plants. Specific plants and trees (e.g. legumes, willow, poplar) can take up contaminants, allowing them to be stored or sorbed within their biomass (e.g. Ref. [141]). The plants are subsequently harvested and either disposed of or incinerated, which creates a waste ash containing the contaminants.
	Monitored natural attenuation (MNA) [142]	Soil and groundwater; predominately hydrocarbons, but 'monitored natural decay' is used to address radionuclides (e.g. caesium-137, strontium-90)	Natural attenuation uses a natural process to reduce the mass, concentration, flux or toxicity of contaminants in soil (or groundwater). MNA requires an assessment to determine the viability of the natural process to achieve the remedial objectives, followed by monitoring and verification to demonstrate that the remedial objectives have been reached. MNA is most effective when the source term has been removed and it is used to address residual contamination. MNA can be utilized as a stand-alone approach or it could be combined with other approaches.

TABLE 5. SUMMARY OF THE COMMON REMEDIAL APPROACHES (cont.)

Approach	Technology	Application	Description and application
Ex situ partial or complete removal	Excavation and disposal	Soil; all types of contaminant	Simple and relatively quick technique involving the excavation of contaminated soil and disposal in an authorized facility. The facility may be on- or off-site and could be a purpose built or repurposed facility. Clean inert fill is needed to backfill the void. There can be an impact to local residents caused by an increase in vehicle movements.
	Detector based segregation [115]	Soil; radionuclides	The approach uses field instruments to support the segregation of contaminant impacted soils. Conveyor belt and bucket monitors can be used as well as handheld monitors. The method assumes that the contaminant fingerprint[a] is consistent throughout impacted soils. Conveyor belt systems can be expensive to set up and therefore this technique is suited to the remediation of large volumes of soil.
	Physical soil washing (dry soil separation) [114, 143]	Soil; predominately heavy metals, including radionuclides	This technique uses particle size separation to separate the finer soil fractions (i.e. silt and clay) from coarse soil fractions. As most contaminants are associated with the finer fractions, by removing these fractions the volume of contaminated soil can be reduced. This approach can be augmented to include detector based segregation of radionuclides.
	Chemical soil washing [143]	Soil; all types of contaminant	This technique either washes out the finer fractions of soil or allows/enables the dissolution of contaminants into solution. The wash water can be modified by adjusting the pH or with the addition of chelating agents, surfactants or leaching agents.

TABLE 5. SUMMARY OF THE COMMON REMEDIAL APPROACHES (cont.)

Approach	Technology	Application	Description and application
	Flotation (also known as flocculation) [114]	Soil; predominately heavy metals, including radionuclides	After removal of the coarse soil fractions, the finer soil fraction is slurried, and an agent is added. Typically, the contaminants are associated with the floating particles. By passing small air bubbles through the slurry, the floating particles collect as a foam on the top, which can be removed. This technique is used extensively in the mining industry to concentrate constituents.
Immobilization or stabilization	Solidification and stabilization [144]	Soil; all types of contaminant	Solidification is the process of fixing the contaminants within a cement to create a low permeability matrix. Stabilization is the use of chemicals to reduce the leachability of contaminants from a soil matrix. Both techniques can be applied in situ or ex situ. This approach is likely to increase the volume of material that requires future management.
	Vitrification [114, 134]	Soil; predominately heavy metals, including radionuclides	Vitrification is the application of high temperature to transform the soil into a glassy matrix. The contaminants are immobilized within the matrix. In situ and ex situ applications are available.

TABLE 5. SUMMARY OF THE COMMON REMEDIAL APPROACHES (cont.)

Approach	Technology	Application	Description and application
Isolation and containment	Barrier systems [115]	Soil and associated groundwater; all types of contaminant	Barrier systems encapsulate the contaminant impacted material and either shield receptors from the contaminants or limit the migration of contaminants via an aqueous or gaseous pathway. Capping layers (horizontal barriers) are the most commonly applied technology. Their construction may be a simple contoured soil cover through to a multilayer system that can include clay, synthetic liners, geomembranes and soil. The cap provides physical separation of the contaminated material and reduces rainwater infiltration. Vertical barriers[b] are used to laterally contain the contaminant impacted material and groundwater. Technologies employed are sheet piles, grout curtains and slurry walls.

TABLE 5. SUMMARY OF THE COMMON REMEDIAL APPROACHES (cont.)

Approach	Technology	Application	Description and application
	Hydraulic barriers [114, 115]	Groundwater; all types of contaminant	Hydraulic barriers can be used to isolate or minimize the movement of a contaminated groundwater plume. A good understanding of the hydrological regime of the site is essential to ensure that these technologies are effective. In addition to vertical barriers (see barrier systems above), there are several technologies that can be applied, such as the following: — Cryogenic barriers: The formation of a frozen barrier to encapsulate contaminants. This approach is technically difficult and costly to apply. It is typically used to control groundwater flow but could be used to isolate a soil source term. — Phytoremediation: Trees can be used for hydraulic control by slowing the movement of contaminated groundwater. The trees act as pumps drawing the water into their roots to stop movement. Electrochemical remediation can be used as an electrokinetic screen to limit radionuclide transport [145].
	Physical treatment [103]	Groundwater; all types of contaminant	These technologies can be used to remove either dissolved or particulate contaminants from groundwater. The mostly commonly used technologies are membrane filtration and aeration.

TABLE 5. SUMMARY OF THE COMMON REMEDIAL APPROACHES (cont.)

Approach	Technology	Application	Description and application
	Chemical treatment [114]	Groundwater; primarily metals, including radionuclides, and inorganic compounds	Chemical separation technologies utilize the chemical properties of the contaminants to remove them from the groundwater. They include the following: — Ion exchange, which removes soluble contaminants through the exchange of contaminants' ions onto a resin. — Chemical precipitation, which removes soluble contaminants from a solution. Equipment can either be batch or continuous flow. — Permeable reactive barriers[c] comprise a treatment agent installed within a subsurface trench that intercepts a contaminant plume [114, 146]. The treatment agents may be sorbents (e.g. activated carbon), chelators and reactive minerals (e.g. limestone).

[a] Also known as vectors or scaling factors.
[b] Also known as impermeable cut-off wall.
[c] Also known as passive treatment walls.

7. ENGAGING INTERESTED PARTIES

An interested party[39] is defined as "A person, company, etc., with a concern or interest in the activities and performance of an organization, business, system, etc." [25]. The need or expectation that interested parties are engaged will vary from country to country and from site to site. In some cases, engaging interested parties will be driven by the site operators, who see the benefits of proactive engagement and working in a transparent manner. For other sites, engaging interested parties will be driven by political or societal demands. Several countries (e.g. USA and UK) include engaging interested parties in key site decisions as a regulatory requirement.

Depending on the location of the legacy trench site, the type of interested party may vary, but in general, they are likely to comprise site employees, the public (individuals and community groups), indigenous and first nation groups, governmental agencies or regulatory bodies (national, regional and local), the media and environmental groups.

During any engagement process to support decision making, it is important to clearly explain the role and responsibilities of the participants. Engagement will enable the needs and aspirations of interested parties to be considered in the outcome of the decision. However, it needs to be clear who the decision maker is, as they are responsible and accountable for the final decision.

The general principles of engaging interested parties will not be covered in this publication, as they are discussed elsewhere [147–155]. This section will highlight the potential concerns that interested parties may have about a legacy trench site and discuss the benefits of engaging interested parties in key decisions for the site.

7.1. EXPECTATIONS AND CONCERNS

Any site where there is or could be an impact to human health or the environment may raise the concerns of interested parties, and attention given to a site may build expectations that some form of remediation may occur.

---

[39] The term 'stakeholder' is used in the same broad sense as 'interested party'. However, the "term *stakeholder* has disputed usage, and it is misleading and too all-encompassing for clear use. In view of the potential for misunderstanding and misrepresentation, use of the term is discouraged in favour of *interested party*" [25].

The following expectations have been commonly raised by interested parties regarding legacy trench sites:

— The site will be cleaned up to remove all waste and contamination and allow any use of the site;
— The next use of the site will bring jobs to the local community;
— There will be jobs for the local community to support the remedial works;
— Funding will be provided to support other community initiatives.

Similarly, the following common concerns have often been raised by interested parties regarding a legacy trench site:

— What is the reason for previous secrecy surrounding the site?
— What might be contained in the legacy trenches?
— Have harmful substances leaked and moved out of the trenches?
— How safe are the legacy trenches in their existing condition?
— How will the decision be made about the future of the legacy trench site?
— How will the views of all the interested parties be included in the decision making process?
— Why is remediation only being considered now, after the site has been there for many years?
— What are the different remediation options available and how will each option affect them?
— How will the best remedial options be selected, and will factors other than cost be taken into account?
— Who will check that the right remedial option has been selected?
— How will residents or the local community be affected whilst the remedial works are taking place? For example, will there be an increase in truck movements?
— If waste is removed, where will it be disposed of or stored?
— Who will fund the remedial works, and will this divert funds from other initiatives that could impact them?
— Who will make sure that the funding is spent appropriately?
— Can the land be reused after the remediation?
— If controls are still needed at the site, who will maintain the controls and how will they be funded?
— If the legacy trenches are not removed, will there be assurance that hazardous materials are not leaving the trenches?

Many of these concerns would also be raised for remediation projects dealing with contaminated land or even with the siting of new radioactive waste disposal sites. A unique challenge for a legacy trench site is to demonstrate that there is enough information to justify either ongoing management or closure of the site with the trenches remaining in situ.

It is important to listen, record and respond to all concerns raised by interested parties to ensure the support for any remedial approach or end state for the site.

7.2. BENEFITS AND CHALLENGES

It is desirable that all interested parties will be positive about the assessment, monitoring and, where appropriate, the remediation of a legacy trench site. Activities to reduce uncertainty and ensure the safety of people and the environment will help to reduce the concerns of interested parties.

Rather than being an obstacle, there are many benefits to undertaking a formalized engagement programme to facilitate dialogue between interested parties and decision makers. Such a programme will provide the site owner, or other responsible party, with the ability to do the following:

- Establish and build trust or rebuild trust where it may have been lost;
- Maintain transparency of the decision making process to avoid invalidation of the decision;
- Minimize the potential for unforeseen issues that may slow or even stop progress at a site (e.g. an interested party requests an independent review of the technical information);
- Build confidence in the decision making process and the final decision to avoid additional work or a decision being overturned;
- Gain the support and perhaps endorsement of the selected remedial options;
- Facilitate the process of risk communication to enable interested parties to make an informed decision;
- In some countries, meet the regulatory requirements regarding the engagement of interested parties.

Despite the positive aspects of engaging interested parties regarding the remediation of legacy trench sites, the process may cause anxiety that did not previously exist, particularly if the local community was not previously aware of the existence of the site. Conducting engagement activities under these circumstances can result in a different set of challenges; for example, the following:

- The local community struggles to accept that the situation may change and prefers to maintain the current situation.

- — Providing answers to the local community in cases where there was no previous awareness of the legacy trench site, especially if there are concerns regarding the performance of the legacy trenches.
- — Explaining that past practices, while not necessarily the same as modern standards, were good practice at the time.
- — Managing lack of understanding and mistrust of the use of environmental modelling. For example, addressing concerns that modelling could be misused to support an option that was considered favourable for other reasons.
- — Rectifying inaccurate or false messages from the past and rebuilding trust.
- — Changing past engagement approaches — which may have occurred as a reaction to a problem — to a proactive approach that seeks to avoid an issue arising.
- — Demonstrating that innovative or optimized approaches may be more practical than purely applying proven technology.
- — Dealing with negative public perception regarding risks from radioactivity. It may also be necessary to explain the potential risks from hazardous waste and/or other contaminants and how these compare with those from the radioactive wastes and/or contaminants.
- — Dealing with concerns about impacts (real or perceived) on human health and the environment.
- — Ensuring that funding is maintained to engage interested parties.

The potential challenges that may be encountered during an engagement programme are not a reason to avoid engaging interested parties. Conducting a dialogue allows both the responsible party and the stakeholders to express their respective views. This approach can help to foster good relationships, although it is important to recognize that this does not guarantee a consensus decision and/or that support for the selected option will be reached quickly. Where there is misunderstanding or a fundamental difference of opinion, engagement can lead to delays and associated cost increases. Therefore, early engagement mitigates these types of risk.

7.3. COMMUNICATION CONSIDERATIONS

Although some concerns from interested parties may be anticipated, others will not be known or understood until the engagement process has started. The communication process encompasses two aspects: the information that the decision maker wishes to communicate and the issues that the interested parties wish to raise and understand. These two aspects need to be brought together within the engagement programme.

The site owner and/or decision maker will ideally communicate the following information regarding the legacy trench site:

— The characteristics of the site, including an understanding of the trench inventory and potential impacts from contamination to people and the environment.
— The uncertainties in the decision making process and how these will be addressed.
— The proposed remedial options and how these have been identified.
— The process of assessing the benefits and detriments of the options to enable the identification of a sustainable and optimized remediation option.
— The possibility that contamination may be left in the ground after the remediation for some or all of the options.
— The benefits and detriments of the options relative to the interested party. These options may have impacts that are either positive or negative, depending on an individual viewpoint. Explaining the impact of no remediation is also important.
— In some instances, the legacy waste trenches may be located in a larger site, which itself is undergoing decommissioning and cleanup. In such a case, it will be necessary to explain how the timing and approach to any remediation of the trenches may fit into the overall site end state strategy.

It may be challenging to explain to interested parties that remaining waste and/or contamination still presents a level of risk, even if it is a very low risk. Risk communication is therefore probably one of the most important aspects of stakeholder engagement and is discussed further in Refs [147, 150, 156].

7.3.1. Communication as part of the decision making process

Current good practice is to assess potential remediation options within a formalized decision making process. A key part of this process is the use of workshops to engage interested parties.

Workshops enable engagement of a range of interested parties at the same time as well as demonstrating transparency in the decision making process. A workshop is an effective way to collect feedback directly from interested parties on the proposed remedial options. In a workshop setting, interested parties can hear about the benefits and detriments of each option and ask questions, raise concerns and give support for an option. A workshop may include an evaluation and scoring process to support direct input from interested parties into the decision.

For a large scale project, weighting and sensitivity factors may be applied to the attributes during the scoring process. In this instance, the inclusion of interested parties in the assessment workshop ensures transparency, builds trust and leads to a greater level of understanding of the final decision. Decision making software (e.g. Hiview3, Expert Choice) can help to provide a visual demonstration of the effects that different weighting and sensitivity can have on the various remediation options, thus allowing an optimized approach to be both understood and supported.

Other approaches that can be used to communicate with interested parties or gain their input include the following:

— Site tours.
— Public meetings.
— Technical forums and training sessions.
— Project information centres. These may be accessible in one location throughout the project or 'pop up' at different locations.
— Questionnaires and opinion surveys.
— Web based meetings.
— Newsletters.
— Web sites.
— Advisory boards or stakeholder group.

Ideally, the person(s) leading the engagement programme will be a qualified communication and engagement expert in the decision maker's organization. When this is not possible or a greater level of independence is needed, personnel from an academic institution or independent consultancy can be used. This is to ensure that technical information is conveyed in a way that is accessible to all interested parties. In some cases, it can be difficult for technical experts to know the best way to present technical information at an appropriate level. Therefore, the skills of a communication expert are essential to ensure that all interested parties can understand the information and make an informed decision.

There are good examples of proactive engagement of interested parties at legacy trench sites. For example, at the Fernald Preserve site, the US DOE Office of Legacy Management's activities include the ongoing communication of groundwater monitoring results and the construction of the Fernald Preserve Visitors Centre (see Ref. [157]). The Fernald Preserve Visitors Centre exhibits tell the history of Fernald Preserve, including the remediation of the legacy trenches, ecological restoration at the site and the ongoing site management activities. The Fernald Preserve Visitors Centre provides an important focal point and educational resource for the local community.

A formal engagement process was used at the Sellafield site in the UK during the assessment of remedial options for the trenches [50]. The process followed the BAT process [102] in order to demonstrate that the most appropriate management strategy was being adopted. The input of interested parties to the assessment process was collected via a workshop. Workshop participants included representatives from the local and regional government offices, the site operator's management team, an independent peer review team, the site stakeholder group and the NDA. Representatives from the Office of Nuclear Regulation and the Environment Agency (EA) attended the workshop as observers; they provided comment and challenge but did not participate in the option scoring exercise.

Further information on communication and consultation with interested parties is provided in IAEA Safety Standards Series No. GSG-6, Communication and Consultation with Interested Parties by the Regulatory Body [151].

8. RECORD MANAGEMENT

In the context of this publication, record management is the collection, organization and maintenance of records. Record management is an essential activity for any site where risks may be posed to people and/or the environment. For legacy trench sites, the need for ongoing good record management is perhaps more apparent, because the existing records are often either inadequate or absent (see Section 4.1).

Record management is essential to support ongoing management of a site and ensure that information is not lost following remediation, for achieving the site end state and after site closure. Where controls are necessary at a site, adequate records contribute to the maintenance of the controls and provide a baseline if any further work is needed at the site (e.g. reports describing remedial activities, records of land use restrictions and monitoring data).

Preserving decision making records is critically important to ensure that confidence is maintained and there is clarity regarding the context and justification of the decision. Without this information, decisions could be undermined or work could be repeated unnecessarily.

Knowledge management[40] is more than just keeping records — it includes managing, sharing and recreating knowledge[41] and can include human resource management and training. In the context of knowledge management, records such as documents, drawings, maps, databases and manuals are termed explicit knowledge.

Guidance on knowledge management, including record management, is well developed, especially in the context of disposal facilities. This section provides a brief overview of the subject, with related discussion specific to legacy trench sites. Further detailed information about knowledge management, including record management, can be found in Refs [23, 24, 32, 158–166].

8.1. DATA MANAGEMENT SYSTEMS

Establishing a data management system can be a major undertaking for any industrial site or waste management site. This can be even more complex for a legacy trench site where the format of the data may have become damaged or obsolete and/or the scope and quality of information collected is different from what is currently needed.

The original format of site records is likely to be diverse and include paper, disks, microfilm, maps, photographs, drawings, electronic files and emails. This information needs to be retained in a formalized data management system so that it can be safely stored, viewed, retrieved and, when necessary, updated. Such a system ensures that the collected information can be maintained for a period commensurate to the agreed site end state or following site. It is also essential that the system provides easy access to the information that states any ongoing institutional control requirements [50].

The system used whilst the site is being actively managed may be different from that used when the site has been closed. For example, a relational database is often used to store, integrate and interrogate the many forms of data, which is useful during active management of a site. If the database is linked with a geographic information system[42], it is easier to visualize and interpret the data. This can support the development of remedial designs and monitoring networks, as well as facilitate engagement with interested parties. Once the site end state

[40] The IAEA defines knowledge management as "An integrated, systematic approach to identifying, managing and sharing an organization's knowledge and enabling groups of people to create new knowledge collectively to help in achieving the organization's objectives" [25].

[41] "The term 'knowledge' is often used to refer to bodies of facts and principles accumulated by humankind over the course of time" [25].

[42] Often referred to as GIS.

has been achieved and/or the site has been closed, records will be moved to an archive for long term storage. Good practice recommends that national archives are maintained by an enduring organization, such as a government body. In the UK, Nucleus is the national nuclear archive, which is being managed by the NDA [167, 168].

8.2. TRANSFER AND MAINTENANCE OF RECORDS

Prior to archiving, or if a national archive does not exist, it can be challenging to define who will keep and maintain records and, if applicable, decide when they may no longer be needed and could be discarded. The custodian of the records could be the original owner or operator of the site, a new owner who has subsequently taken over responsibility, a nationally appointed custodian or a governmental department.

Table 6 identifies key activities or decision points during the lifetime of a legacy trench site and highlights potential questions and the associated information that needs to be maintained. The research conducted by the Australian Nuclear Science and Technology Organisation (ANSTO) for the LFLS particularly highlights the importance of records made prior to the start of operations to provide a baseline of the environmental characteristics of the site [169, 170].

TABLE 6. KEY ACTIVITIES OR DECISION POINTS IN THE LIFE CYCLE OF A LEGACY TRENCH SITE

Activity or decision point	Potential question	Information needed
Site selection and characterization	Why was the site chosen? What was the original status of the site?	Historical records, aerial photographs, institutional files, site selection decisions, baseline environmental report.
Operation of legacy trench site	What was the operational period? What standards or acceptance criteria were used?	Records of type of disposal, volumes, inventory, dates of disposal and locations within the trench footprint area.
Cessation of site activities	Did disposals cease after nuclear related site activities were complete?	Records of trench management during any decommissioning activities.

TABLE 6. KEY ACTIVITIES OR DECISION POINTS IN THE LIFE CYCLE OF A LEGACY TRENCH SITE (cont.)

Activity or decision point	Potential question	Information needed
Change of trench facility operator	Did trench disposals continue?	Evidence of the transfer of records and knowledge and their continued maintenance.
Control period of the site	Did the site undergo a period of control?	Evidence of the transfer of records and knowledge to a potential new custodian. Records related to environmental monitoring.
Post-operation decision making	Why did disposals cease? What management and remediation decisions were made upon site closure?	Records that define the reasons and decision to stop. Records of proposed options and decisions.
Site characterization	Was there a site characterization programme?	Records of all site characterization data.
Management and remediation	What, if any, management and remediation approaches were adopted for the trenches?	Records and knowledge of the adopted option and how it was deployed.
Post-remediation monitoring	Has there been any post-remediation monitoring?	Records of monitoring programme.
Stakeholder engagement	Was there a stakeholder engagement programme?	Evidence of stakeholder concerns and whether they were considered in the decisions made.

Common causes of data loss include the update of data storage technology, intentional or accidental deletion of electronic files, misplacement of physical records and damage to storage media. These issues can be prevented by storing duplicates at two or more physical sites and/or utilizing different storage media. The transfer of records from one organization to another, including from an operator to the regulatory body, can also create the risk of records being lost. Careful planning and validation are needed to ensure that information is not lost during the transfer process. It is important that the potential for record loss is acknowledged and appropriate mitigation measures are put in place.

9. CONCLUSIONS

Legacy trench sites present a unique group of contaminated land sites. Legacy trench sites can pose an additional challenge compared with other contaminated land sites because of the heterogeneous nature of the disposed waste. These characteristics create challenges throughout the life cycle of the site, especially during site investigation, environmental modelling and remedial activities.

Uncertainties regarding the inventory, construction and operation of the trenches requires ingenuity and dedication to ensure that the potential risks to people and the environment are understood. Experience at legacy trench sites clearly demonstrates the importance of inventory records. Understanding the inventory is key to assessing the potential risks posed by the site. Where records exist for legacy trench sites, it is unlikely that they meet the documentation requirements of a modern disposal facility. Multiple record sources can be used to create an inventory for the site. However, a key learning point is that even if available records do not mention a contaminant, the presence of the contaminant cannot necessarily be ruled out. Additional investigative work may be necessary, such as an understanding of the operational practices of facilities sending waste to the trenches. For example, it appears to have been common practice at some sites not to record tritium on the records although monitoring has confirmed that it is present.

Establishing the inventory is likely to require extensive resources. Even with this work, uncertainties can remain owing to missing, incomplete or poor quality records. A balanced decision is necessary to decide how much resource will be used to establish the inventory or whether the uncertainty can be managed in other cost effective ways.

Site investigation can be challenging at any contaminated land site, especially where multiple past practices were employed, leading to additional complexities. The benefit of obtaining additional sampling data needs to be balanced against the risk to current workers and the potential risk of creating conduits, cross-contamination or contaminant mobilization. It is good practice to consider non-intrusive methods as part of the site characterization at a legacy trench site.

Caution needs to be exercised in the conservatism used in risk assessment, which may lead to an overprediction of the impact that the site poses to people and/or the environment. This can subsequently lead to unnecessary remediation activities. It is essential that the principle of optimization is used to ensure that a balanced approach is taken for the end state decision.

The experience described in this publication illustrates that, just as for other contaminated land sites, there is no 'one size fits all' solution for legacy trench sites. Management and remediation strategies can vary from long term monitoring to ensure natural attenuation, to waste retrieval, depending on the physical characteristics of the site (e.g. inventory, hydrogeology) and the context of the site (e.g. national strategy, societal-economic context, attitude of interested parties).

International experience on the management and remediation of legacy trench sites continues to grow. A particular challenge, which is not unique to legacy trench sites, is how to define the site end state. The site end state comprises the physical state, the controls (where necessary) and the potential uses. The decision making process is iterative and needs to be clearly recorded to allow decisions to be reviewed and understood in the future. Aside from the technical challenges, social and economic factors have a great influence on the outcome. Decision making can take place only when the uncertainties have been reduced to meet technical requirements and ensure stakeholder confidence. A strategic site end state and/or interim states may be used to enable progress at the site.

It is important to recognize that legacy trench sites exist in many countries around the world and there needs to be continued engagement in the international community to strengthen the approach and resources to continue to manage these sites safely.

Appendix I

CASE STUDIES

I.1. ARGENTINA, EZEIZA RADIOACTIVE WASTE MANAGEMENT AREA

I.1.1. Site setting and description

The National Atomic Energy Commission of Argentina (CNEA) was created in 1950 to oversee the development of applications for the peaceful use of nuclear energy. The scope of CNEA includes the management of LLWs, research and development, production of radioisotopes for medical and industrial applications, uranium mining and processing, fuel element manufacture and the operation of three nuclear power plants.

CAE was created as a nuclear research facility and covers approximately 830 hectares (one hectare is 10^4 m^2). The site is located approximately 35 km from the city of Buenos Aires and close to the Ezeiza International Airport. The site comprises research reactors, production plants for radioisotopes and nuclear fuel, and a facility for the management and storage of spent fuel and other radioactive waste.

AGE covers eight hectares, and forms part of CAE. AGE began operation in 1969 and carries out waste management activities including treatment, conditioning and final disposal of low level solid and liquid radioactive waste. AGE comprises a semi-containment system for solid radioactive waste, a semi-containment system for very low level and very short lived liquid radioactive waste, two deep underground silos for structural radioactive wastes and sealed sources, and other waste handling and storage facilities (see Fig. 20). In 2001, CNEA decided to conclude all permanent disposal operations at the site. AGE continues to be used for the storage of conditioned low and intermediate level radioactive wastes, awaiting the construction of an appropriate repository.

I.1.2. Operational history of the site

The Semi-containment System for Solid Radioactive Waste comprises two trenches. The term ‘semi-containment’ is related to the assumption that, because of its design, a gradual natural degradation of the barriers would occur, resulting in the progressive release of contaminants over time. This approach to waste facility design was common practice in the 1960s. The trenches contain waste

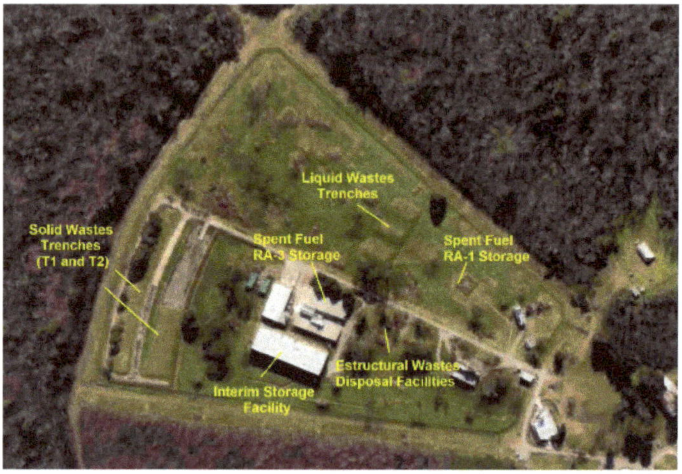

FIG. 20. Site layout of Área de Gestión Ezeiza (Ezeiza Radioactive Waste Management Area). Figure courtesy of División Área Gestión Ezeiza.

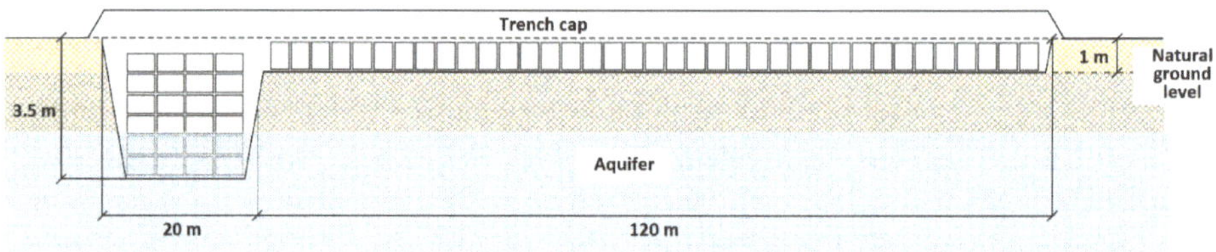

FIG. 21. Schematic of the longitudinal section of the T1 trench, showing the placement of the drums.

from nuclear power plants, research and medicine centres, and industry sites across Argentina.

Trench No. 1 (T1 trench) was commissioned in 1969; the trench is 140 m long and 10 m wide and contains 3400 conditioned waste drums. The trench has no engineering and was initially excavated to 2.5 m below natural ground level. The drums were placed on their sides on top of each other in the trench. Repeated flooding events occurred during the initial operation owing to rising water level in the underlying aquifer. Therefore, the excavated trench could no longer be used, and the drums were stacked vertically to a height of 1 m above the natural ground level (Fig. 21).

In 1980, a cover was applied to the first part of the T1 trench, and complete closure occurred in 1988.

FIG. 22. Photograph showing the Semi-containment System for Solid Radioactive Waste at Área de Gestión Ezeiza. Trench No. 1 (T1) is covered and Trench No. 2 (T2) is partially covered. Photograph courtesy of División Área Gestión Ezeiza.

Trench No. 2 (T2 trench) was commissioned in 1989 and operations ceased in 1999. The trench is 120 m long and 20 m wide and has a capacity of 5600 drums. The trench was constructed at ground level onto compacted calcareous silty soil with a levelled broken stone bed and a 30 cm concrete perimeter retaining wall. Approximately a third of the trench was filled and covered, and the rest of the trench remains uncovered. Between 1995 and 1999, drums continued to be placed in the uncovered portion of the T2 trench.

The covers for the last section of the T1 trench and the first third of the T2 trench both comprisedcompacted calcareous silty soil formed into a mound. The mound was covered with a bituminous material, a layer of sand, a 200 μm polyethylene sheet and another layer of compacted calcareous silty soil. Finally, the mounds were covered with soil and seeded with grass (Fig. 22).

CNEA suspended solid disposals to AGE in 1999, and in 2001 all waste disposal operations ceased. Leading up to these decisions, it was recognized that there were changes in the environmental and socioeconomic context of the site. An increase in annual rainfall and the artesian nature of the Puelche aquifer brought the local groundwater aquifer closer to ground level, potentially enabling greater contact with the waste facilities. Social and economic demands resulted in the development of land close to AGE [171].

In 2000, there were concerns from interested parties that the site could potentially be contaminating the underground drinking water sources. A court proceeding took place and CNEA was required to undertake an environmental study on AGE with independent assessment from the Nuclear Regulatory Authority (ARN).

Degradation of the uncovered drums in the T2 trench was observed as they remained exposed to the harsh weather conditions. In 2004, ARN authorized closure of the T2 trench, with specific requirements defined for the uncovered section of the trench. However, in 2005 the judge overseeing the court proceedings ordered the removal of the drums in the uncovered section of the T2 trench. In 2010, these drums were conditioned and packed into overseas containers and moved to the Long Term Storage Deposit. The Long Term Storage Deposit is located outside AGE but within CAE and was especially built for waste storage until the construction of a new waste repository [172, 173].

Extensive site investigation, monitoring and environmental modelling were carried out between 2003 and 2007. These works were supported by the IAEA and also via a technical assistance project with the US DOE[43].

I.1.3. Regulatory context

In 1994, through Decree 1540, a national nuclear regulatory body was created, ARN, making the regulatory functions independent. The T1 trench was operated without a licence (from 1969 to 1988) and the wastes contained in the trench are considered to be historic wastes [174]. The T2 trench was operated between 1989 and 1995 without a licence.

In 1995, ARN issued AGE an operating licence, for which it required the responsible entity, CNEA, to present the corresponding mandatory documentation. All wastes disposed of prior to 1995 are considered to be historic wastes [174].

I.1.4. Drivers for site assessment

In 1999, the driver of the assessment was the closure of the T1 trench and the T2 trench, as both trenches had completed their operational phases. At the time, CNEA had recognized the changes in the environmental conditions and socioeconomic context of the site and hence identified that further safety assessment of the site was required.

[43] The Site Characterization, Monitoring and Modelling project was started in 2003 and sponsored by the Joint Coordinating Committee for Radioactive and Mixed Waste Management (JCCRM) within the framework of the Technological and Scientific Cooperation Agreement between CNEA and US DOE, with the collaboration of Florida State University [172].

I.1.5. Site evaluation

During the initial evaluation stage, limited information was available regarding the construction of the T1 trench and the waste inventory. To support the safety assessment, the trench design and conservative estimates of the inventory were prepared from data archived in the AGE database, handwritten daily reports and historic maps and photographs. This information was supported through interviews with the former and current employees at the site.

The major site investigation works were carried out during 2003 to 2007. The activities included the following:

- Geophysical surveys, including ground penetrating radar, vertical electrical sounding resistivity surveying and electro-magnetometry;
- Borehole drilling to establish geological and hydrogeological ground conditions, collection of samples and installation of groundwater monitoring boreholes;
- Hydrogeological studies, including pumping tests;
- Soil and sediment analyses, including contaminant concentrations and groundwater flow and transport parameters;
- Groundwater monitoring, including the installation of continuous phreatic level meters;
- Groundwater sample analyses;
- Estimations of distribution coefficients (K_d) of different radionuclides of interest (uranium, strontium, cobalt, caesium and plutonium) using sediments collected from the site.

The safety assessment was undertaken using the ISAM methodology [65]. The objective of the assessment was to determine the radiological consequences of the semi-containment systems for solid wastes for the post-closure period in the long term. The assessment was applied to normal and intrusion release scenarios with an initial assumption of a 50 year institutional control period. The interpretation of the results of the assessment suggested that the period of institutional control needs to be longer than 50 years, primarily owing to the human intrusion scenario.

A hydrogeological conceptual model has been produced for the site to create a flow and transport model. The code Visual MODFLOW (Modular Three-Dimensional Finite-Difference Groundwater Flow) 4.1 was used, which allows the modelling of a three dimensional aquifer system. This enabled the calculation of the concentration of the chemical species dissolved in an aquifer within the modelled area covering AGE and the surrounding area.

A dosimetric model was used to relate the concentration of radionuclides with the doses received by the exposure group from the ingestion of drinking water. These wells were located between the T1 trench and the Aguirre stream. The model was run to identify when the peak dose to the critical group would occur.

The safety assessment was completed and issued to ARN in 2007. ARN did not accept this assessment, because it considered that the radiological inventory could be improved and optimized. Additionally, ARN requested an increase of the radionuclides considered in the inventory from 10 to 24. The additional radionuclides included several radionuclides from nuclear power plants and four uranium isotopes added by the National Radioactive Waste Management Programme. With this objective in mind, a radiological characterization plan of the waste was carried out to obtain a new inventory.

Meanwhile, in agreement with ARN, a second safety assessment was carried out to evaluate the site during normal operation. The two events that were considered owing to their probability of occurrence were a flooding disruption scenario and an aircraft crash disruption scenario. The assessment was carried out using a conservative preliminary inventory. In 2012, ARN approved this second safety assessment of the solid waste semi-containment systems (T1 and T2 trenches). As yet, no physical works have yet been carried out to close the site. Monitoring and improvements to the environmental modelling and safety assessment are ongoing. The normal release and human intrusion scenarios will be considered in the definitive safety reassessment when the full radiological inventory is completed.

In recent years, the radiological characterization plan has focused on the characterization of the drums from the uncovered section of the T2 trench to improve the inventory estimations for the safety assessments. One hundred and fourteen drums were selected for radiological characterization from the 1789 drums that were conditioned and moved to the Long Term Storage Deposit. The drums have been gamma scanned and sampled to allow scaling factors to be developed.

I.1.6. Anomalies present in trenches T1 and T2 during their operation

The results of the monitoring carried out since the start of the environmental characterization of AGE indicated the presence of tritium and uranium in the aquifer downstream of the T1 trench. It was assumed that these contaminants came from the deeper southern section of the T1 trench, where the waste drums are expected to be in contact with the groundwater. The plume is limited to a localized sector downstream of AGE and is within the CAE site boundary. Access to this area is restricted, as it forms part of CAE and hence it cannot be accessed by members of the public.

The monitoring results show that in the direction of groundwater flow, there is a decrease in the concentration of tritium and uranium. Figure 23 shows the modelled tritium plume overlaid on the aerial satellite image of AGE and the surrounding area and Fig. 24 shows the decrease in tritium concentrations with distance from the trenches.

In recent years, there has been a decrease in the concentration of tritium in monitoring well T5 but some increases in tritium concentrations in monitoring wells T4, T3, T2 and T1. Uranium is present but is now found only in monitoring well T5.

The monitoring data show the release of radionuclides at irregular intervals with a dependence on the oscillations of the groundwater level of the aquifer. In general, the highest levels of tritium occur together with rises in the groundwater level. The increases in groundwater level cause movement of the water and consequently the transport of tritium. Hence, tritium enters the groundwater in the form of 'pulses'. The tritium activity can vary up to three orders of magnitude.

The concentration of tritium in the Aguirre stream has always been below the limits of detection. The Aguirre stream is located 385 m down the hydraulic gradient from the trenches and is a natural discharge point for groundwater.

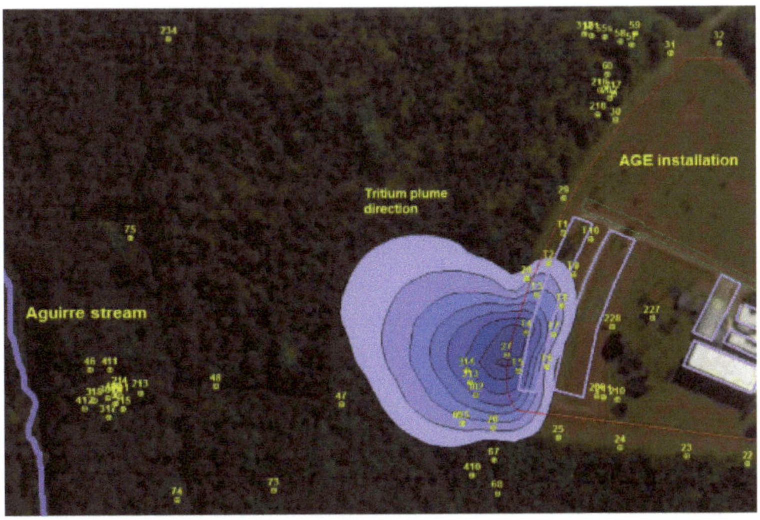

FIG. 23. Modelled tritium plume associated with the Semi-containment System for Solid Radioactive Waste (Trench No. 1). Groundwater samples were collected on 21 July 2016. Figure courtesy of División Área Gestión Ezeiza.

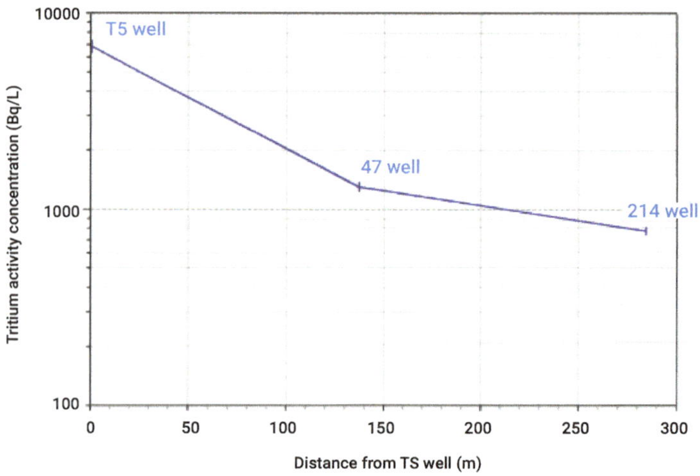

FIG. 24. Decreased concentration of tritium activity with distance, measured with the Semi-containment System for Solid Radioactive Waste (Trench No. 1). Groundwater samples were collected on 15 May 2018. Figure courtesy of División Área Gestión Ezeiza.

I.1.7. Remediation options and selected approach

Environmental studies at the site have verified that monitored natural attenuation is a viable option for remediation. It is considered the most sustainable option and the least invasive method for environmental remediation.

Depending on the conclusion of the current revision of the safety assessment and ongoing monitoring, other remedial options may be considered. These options would focus on improving the cover system of the trenches and addressing the contaminant impact from the drums located in the deeper zone of the T1 trench. The selected option will depend on a range of factors, including the safety of the public and workers, technical feasibility and financial costs.

The option assessment will consider the full range of options, including monitored natural attenuation, placement of a capillary barrier or intrusion resistant cover, and excavation of the drums from the deepest section of the T1 trench.

I.1.8. Implementation of the remediation approach

The implementation of remediation measures has not yet been confirmed. Groundwater monitoring and evaluation continues at the site as part of normal operations.

The sector between AGE and the Aguirre stream is part of the area's natural hydrogeological system. There remains some uncertainty about the evolution of this system in the long term, especially if there is a change in future land use. For this reason, in 2015 the institutional control area for AGE was extended from the original site boundary to the Aguirre stream. Therefore, all the institutional control measures for AGE have been extended to this area, including additional physical security, radiological protection and routine environmental monitoring.

I.1.9. Conclusions and lessons learned

AGE is typical of many legacy trench sites in that its location was based on operational needs rather than identifying a suitable location on the basis of the geological and hydrogeological setting. In addition, not enough forethought was given to the potential expansion of Buenos Aires.

In the past 20 years, CNEA has undertaken many studies to underpin the safety case for the operation and closure of AGE. This demonstrates the extent of work that may be required to demonstrate that historic waste disposal sites do not pose a risk to people or the environment.

A key challenge has been addressing updates to the regulatory requirements as safety standards evolve. This demonstrates the value of maintaining good working relationships with the regulators.

I.2. AUSTRALIA, LITTLE FOREST LEGACY SITE

I.2.1. Site setting and description

ANSTO is a statutory body of the Australian government and was created in 1987 to replace the Australian Atomic Energy Commission. ANSTO's nuclear research site is located at the Lucas Heights campus. The site includes the operational Open-Pool Australian Lightwater (OPAL) research reactor, neutron scattering instruments, research laboratories and the High Flux Australian Reactor (HIFAR), which is currently being decommissioned. ANSTO is also responsible for the LFLS, which comprises trenches that were used for the disposal of radioactive waste between 1960 and 1968.

The LFLS is located near the Lucas Heights campus, on the southern edge of the Sydney metropolitan area in Australia (see Fig. 25 (a)).

ANSTO has carried out ongoing care, maintenance, surveillance and monitoring at the LFLS. In addition, ANSTO has undertaken research activities at the LFLS that have contributed to international research on legacy trench sites.

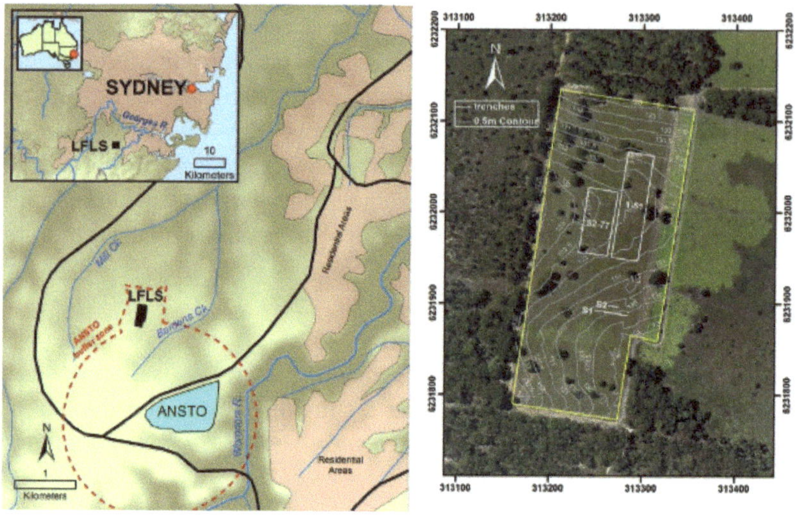

(a) (b)

FIG. 25. (a) Location of the Little Forest Legacy Site. (b) Aerial view showing the trenched areas and the site boundary. Image courtesy of the Australian Nuclear Science and Technology Organisation.

In recent years, a detailed research project has been implemented at the LFLS, which has been reported in several publications (e.g. Refs [21, 27, 175–182]).

The LFLS is situated inside the ANSTO buffer zone. A general introduction to the site can be found in Ref. [21]. The site occupies a section of land extending just beyond the 1.6 km radius circle (buffer zone) around the former HIFAR at Lucas Heights (Fig. 25 (a)). The suburb of Barden Ridge, located 2.5 km to the east, is the nearest residential area to the LFLS, with the western parts of the suburb of Menai approximately 3 km to the north-east. Future suburban expansion and other developments in the general area surrounding the LFLS are expected.

The site is mainly covered by grass, which is mown on a regular basis. The vegetation of the surrounding area is mostly native shale forest. The topography of the site is gently sloping. Two sets of trenches containing the wastes occupy the higher part of the site (see Fig. 25 (b) and Fig. 26). Two additional trenches (S1 and S2) are located about 50 m to the south of the main trenched areas. These were filled in 1967 (S1) and 1968 (S2).

The trenches were excavated in the shallow soil layers above a shale lens, through which infiltration is limited, acting as a partial barrier to direct downward movement of groundwater into the aquifer in the Hawkesbury Sandstone below. There is some intermittent groundwater seepage at the shale–sandstone interface that outcrops south-east of the LFLS [175]. The upper sandstone units of the Hawkesbury sandstone are generally weathered and strongly jointed and fractured.

The LFLS is adjacent to several other waste disposal sites, including (on the western boundary) a former municipal waste disposal site. Further west is the Lucas Heights Waste Management Centre, which operates a major municipal landfill. Immediately to the east of the LFLS is another impacted area, which was formerly used for the disposal of night soil (sewage). A former industrial liquid waste site used for the disposal of chemicals, including paints, grease, solvents, tannery wastes, as well as specific hazardous industrial chemicals including dioxin contaminated materials and residues from herbicide production, is located to the north-west.

I.2.2. Operational history of the site

As shown in Fig. 26, the trenches were filled sequentially from 1960 to 1968. The trenches were approximately 25 m long, 0.6 m wide and 3 m deep, and spaced 2.7 m apart, with the wastes covered by about 1 m of the local clay soil.

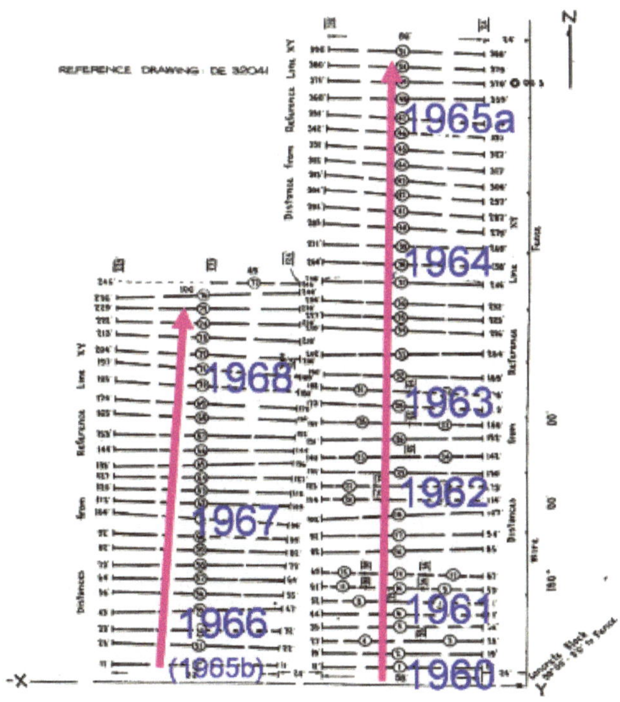

FIG. 26. Sequence of filling of the trenches at the Little Forest Legacy Site from 1960 to 1968. Image courtesy of the Australian Nuclear Science and Technology Organisation.

Records were kept of the disposal operations, providing a general indication of the contents of each trench. Further information is given in Ref. [21].

The waste contained various radionuclides, including tritium, activation products such as colbalt-60 (which was used in an irradiation facility at the site) and fission products (caesium-137 and strontium-90). Actinides were also disposed of, including uranium-238, uranium-235 and thorium-232 and their decay products, and small (several grams) amounts of plutonium-239, plutonium-240 and uranium-233. The wastes were generated from the research into power reactor design that was being undertaken at the time. This research activity also resulted in significant amounts of beryllium (approximately 1100 kg) being disposed of into the LFLS trenches.

The wastes consisted of a variety of waste from the ANSTO laboratories and waste packages consigned from other organizations, including radioactive materials, contaminated equipment and beryllium and/or beryllium oxide scrap [175]. Numerous containers of liquid waste were either being buried or emptied into the trenches. Some items were incinerated on the site. Amongst the items disposed of at LFLS were 760 drums of solidified sludge from ANSTO's effluent treatment plant. Further details of the wastes and estimates of the

FIG. 27. Waste materials emplaced in trenches at the Little Forest Legacy Site during disposal operations. Photographs courtesy of the Australian Nuclear Science and Technology Organisation.

inventory are given in various sources [21, 175]. Figure 27 shows waste materials placed into the trenches before backfilling occurred.

I.2.3. Regulatory context

Australia is governed by a Federal (Commonwealth) Government and has several individual States and Territories. ANSTO is a Commonwealth Government Agency and is subject to Commonwealth, rather than State, statutory legislation. The Australian Radiation Protection and Nuclear Safety Act 1998 and supporting Regulations are specifically relevant to ANSTO operations and the LFLS. The Act is regulated by the Australian Radiation Protection and Nuclear Safety Agency (ARPANSA), an independent Commonwealth Authority.

The LFLS was the first site to be licensed as a legacy site under the Australian Radiation Protection and Nuclear Safety Act 1998. This occurred in July 2016, following amendments to the Australian Radiation Protection and Nuclear Safety Act. A condition was placed on this licence that required ANSTO to develop a plan for managing the wastes and the facility over the medium term (one to two decades) and subsequently over the longer term.

I.2.4. Drivers for site assessment

As highlighted above, the site was licensed as a legacy site in 2016 and ANSTO was required to develop a plan outlining the arrangements for the future management of the facility and its wastes.

For several years, a research project aimed at fully characterizing the site and providing input to future management and possible remediation decisions has been underway at the LFLS. The rationale and background of this project is described in Ref. [21].

I.2.5. Site evaluation

To understand the waste inventory, there has been a great effort to review historic records. The types of information available regarding waste disposals at the LFLS include waste disposal cards (‘pink cards’), waste burial books and photographs (see Figs 11 and 12). The pink cards contain descriptions of items that were generated by the originator of the wastes. However, the available pink cards cover only a small proportion of the approximately 50 000 disposed items [176]. Although the information shown in the waste burial record is less detailed than the pink cards, it contains entries for the majority of items disposed of at the LFLS.

The LFLS is being extensively characterized by a research project coordinated by ANSTO and involving several other participants, including research groups at several universities. Research outcomes include journal papers on various topics, including the mobility of tritium [178], a general description of the bath-tubbing effect and the dispersion of radionuclides to the surface [177], the groundwater geochemistry [179] and possible mechanisms causing plutonium mobilization [180]. Other recent papers discussed the possible roles of microorganisms [181] and co-disposed organic chemicals, such as tri-butyl phosphate [27], in radionuclide mobilization at the site.

A qualitative description of the bath-tubbing process was presented in Ref. [177], where it was shown that the measured water level in one of the legacy trenches responded rapidly to rainfall. The level could rise to the surface during intense rainfall events. Figure 28 shows the conceptual site model; the main feature is water overflowing at the end of the trenches during times of high water levels. Since the trench water contains radionuclides such as plutonium and americium, this has resulted in these radionuclides accumulating in the shallow soil layers and on the ground surface. The fluctuating geochemical conditions play a key role in mobilizing radionuclides from the trenches [182], but the hydrology of the site is still under investigation.

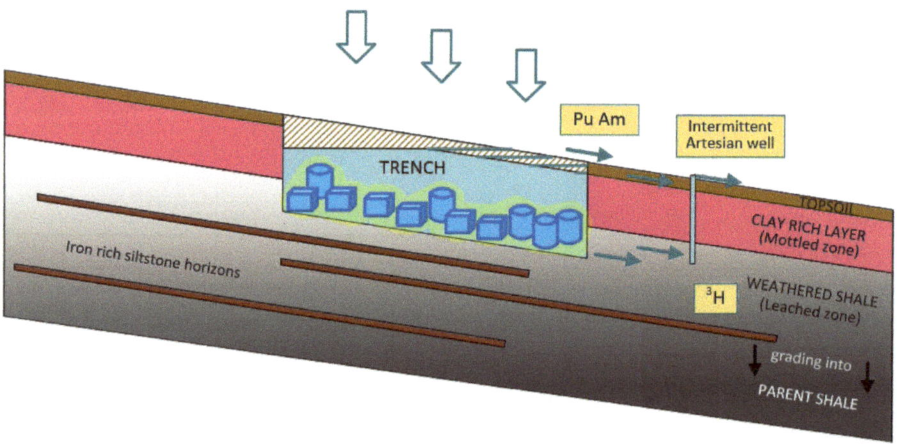

FIG. 28. Conceptual model for the trench bath-tubbing scenario for the Little Forest Legacy Site. Reproduced from a figure provided courtesy of the Australian Nuclear Science and Technology Organisation.

I.2.6. Remediation options and selected approach

As discussed above, the licence conditions as defined by ARPANSA include a requirement for a site management plan, with options for the site end state that may include remediation. The current ANSTO projects include characterization of the site and assessment of available in situ remediation options, which may be more cost effective in comparison with exhumation of the waste. In this project, a test trench facility is being constructed to support the in-field evaluation of remediation options, facilitate detailed studies of bath-tubbing processes and enable model parameterization.

The remediation objectives and preferred management option have not been decided for the site. Several options are under consideration, including the following:

- — Ongoing passive management and monitoring;
- — In situ remediation, such as a geo-engineered cover;
- — Decontamination of trench water by 'pump and treat' methods;
- — Stabilization of trench contents by in situ grouting;
- — Exhumation of the trenches and possibly part of the surrounding area and disposal in another site.

Further work is needed to bound the contextual uncertainties associated with the decision, for example, the requirements of the national radioactive waste strategy.

I.2.7. Implementation of the remediation approach

The site is still being characterized and modelled. A range of potential remediation options are being assessed.

I.2.8. Conclusions and lessons learned

The costs and benefits of the potential options are presently being evaluated as part of ANSTO's investigations of the site. As part of this work, an experimental trench facility in the vicinity of the legacy trenches has been constructed. Future activities will be focused on the experimental trenches, including hydrological studies and tracer tests, as well as studies of engineering interventions and in-trench chemical processes.

I.3. CANADA, CHALK RIVER LABORATORIES

I.3.1. Site setting and description

The Canadian Nuclear Laboratories Chalk River Laboratories (CRL) site is located in Renfrew County, Province of Ontario, on the south shore of the Ottawa River (Fig. 29).

Land use in the region consists primarily of forestry, recreation and tourism, with limited agriculture, trapping and mining. The area supports a wide range of wildlife species, including various species at risk. The site consists of gently rolling hills made of a mixture of exposed bedrock, glacial till, fluvial sand interspersed with small lakes and marshes [183].

The CRL site itself covers 40 km^2 and is divided into a 'built-up area' adjacent to the Ottawa River and a 'supervised area' which comprises the remaining, mainly wooded, property to the west (Fig. 30). The built-up area contains reactors, laboratory buildings and other site support facilities. The waste management areas are located in the supervised area.

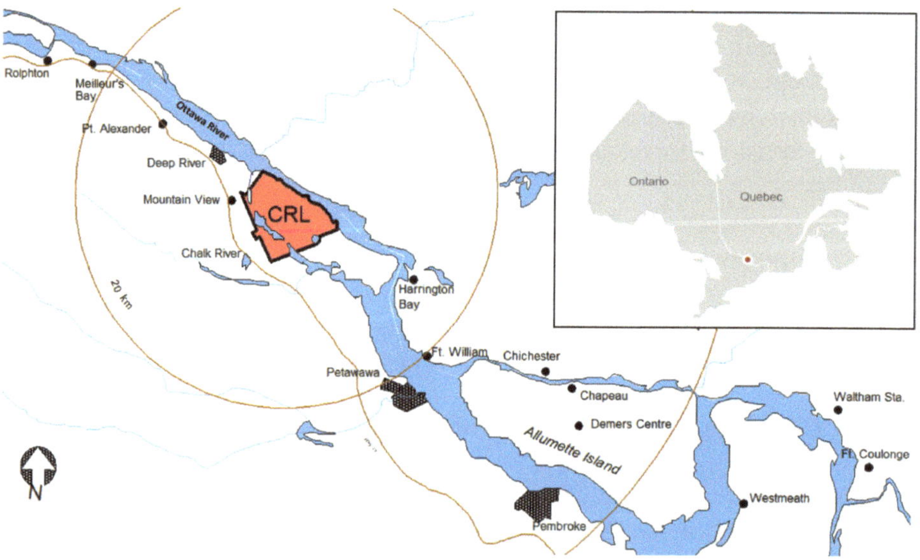

FIG. 29. Location of Chalk River Laboratories site. Figure courtesy of Atomic Energy of Canada Ltd.

FIG. 30. Aerial photograph of the Chalk River Laboratories site looking north-west. The Ottawa River and built-up area are in the foreground and the supervised area is in the distance. Figure courtesy of Atomic Energy of Canada Ltd.

I.3.2. Operational history of the site

Construction at CRL began in August 1944. Nuclear research at the Chalk River site began shortly afterwards in the Zero Energy Experimental Pile (ZEEP) and National Research Experimental (NRX) reactor and other nuclear research laboratory buildings. Support facilities such as analytical laboratories, engineering workshops and stores, and services such as administration, radiation protection, environmental and biological research, nuclear materials and waste management were created, as needed.

In 1952, the NRX reactor suffered an accident that resulted in extensive fuel failure, severe damage to the core and release of radioactive material. Solid and liquid wastes, including reactor components, were taken to the waste management areas. In 1954, the NRX reactor was returned to service.

The research focus shifted in 1954 from the production and recovery of plutonium and uranium to the application of nuclear technology for isotope production and electrical power generation based on the concept of the natural uranium fuelled, heavy water moderated CANDU (Canada deuterium–uranium) reactor. Various facilities were installed to support this new mission. The National Research Universal (NRU) reactor was brought on-line in 1957 and was shut down in 2018.

I.3.2.1. *Waste management areas*

Beginning in 1946, wastes generated at the CRL site were placed in unlined trenches in the sand deposits in the supervised area and then covered with additional sand to protect workers and prevent contamination from being dispersed by the wind. In 1952, the NRX reactor accident resulted in 4500 m^3 of cooling water being diverted to the sand trenches in the initial waste management area. As a result, this earliest waste management area was no longer fit for its original purpose, and a second waste management area was developed 750 m farther to the west. The locations of various waste management areas are shown in Fig. 31.

In 1956, additional engineered facilities, developed for specific types of waste, were added with varying engineered containment. LLW continued to be placed in trenches dug in the sand. ILW was placed in asphalt trenches to provide some degree of isolation from the environment. Higher level and longer lived waste was placed in 'tile holes' (concrete drainage tiles placed vertically below

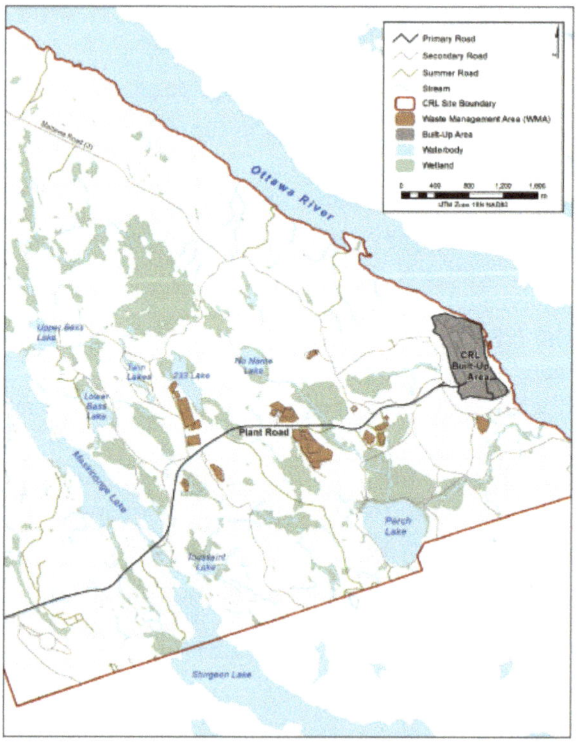

FIG. 31. Chalk River Laboratories site layout including waste management areas. Figure courtesy of Atomic Energy of Canada Ltd.

the ground surface on a concrete base) to provide isolation and shielding for workers. By 1959, the cap on one of the asphalt trenches for ILW had failed, leading to the practice of using concrete for the trenches instead of asphalt. Brief descriptions of each type of trench follow.

Sand trenches

Sand trenches were dug approximately 3 m deep and 2 m wide in the superficial sand deposits that cover much of the CRL site. Waste was placed directly into the trench, which was then backfilled with sand. Both solid and liquid wastes were placed in the trenches, although the liquid wastes were generally in containers at the time of emplacement. Figure 32 shows a typical waste trench.

Asphalt trenches

Reports from the 1950s indicate that a 1.8 m deep trench was dug in the sand. A wooden box was built in the trench and asphalt was poured to form a floor. Plywood sides were erected about 7.5 cm to 10 cm inside the wooden box and the space between the wooden box and the plywood sides was filled with asphalt. Wastes were placed inside, with sand filling the interstices. Asphalt was poured on top to form a roof. The trench was then covered with sand. The failure of one of the asphalt trenches in 1959 led to the change to concrete construction.

FIG. 32. Typical unlined waste trench with waste dug in the sand at the Chalk River Laboratories site. Photograph courtesy of Atomic Energy of Canada Ltd.

FIG. 33. Single width concrete waste trench constructed at the Chalk River Laboratories site. Left: Empty. Right: with emplaced waste. Photographs courtesy of Atomic Energy of Canada Ltd.

Concrete trenches

The first two concrete waste trenches were constructed of poured concrete and were 1.8 m by 1.8 m by 60 m; the dimensions of the third trench were 2.4 m by 2.4 m by 60 m. Each trench was divided into 12 m sections. Figure 33 shows the first concrete trench prior to filling (left) and after filling, just prior to subsequent burial (right). Later trenches were built to be twice as wide.

I.3.3. Records description

Historical records play an important role in environmental remediation at the CRL site. They provide input for hazard and risk assessments and contribute to various activities, including the following:

- Making end state decisions;
- Planning remediation project technical details, such as remote or contact handling waste retrieval methods;
- Ensuring worker safety during implementation of the remediation project;
- Determining waste categories and volumes for post-project waste disposition pathways.

Historical records at CRL include the following:

— Disposal slips. These records were filled out by the waste generator, the person who, presumably, would have had the greatest and most accurate knowledge of the waste package and what it contained. Figure 34 shows an example of a Waste Disposal Slip.

— Waste management area logbooks. Logbooks were kept by waste management area personnel. Waste packages were documented as they were received by the waste management area for disposal.

— Monthly reports to Health Canada. The waste management area supervisor completed a monthly summary report of the amount of activity deposited in a given facility. This report was sent to Health Canada (regulator) (as opposed to the Atomic Energy Control Board — the name of the nuclear regulator at the time).

— Internal quarterly division progress reports (e.g. Engineering Division, Biology Division, Maintenance and Construction Division).

— Environmental Panel Reports. Reports of oversight bodies have descriptions of waste management practices of the day. If approval for a particular practice was given, the conditions governing that practice were also given.

— Memoranda and miscellaneous correspondence. Historical memos and letters have documented waste management practices that were used in the past (e.g. the timing or location of 'special burials'). Occasionally, larger projects required special permission for one-off special disposals. For example, internal correspondence between the head of the Operations Branch and the head of the Waste Disposal Branch would give an indication

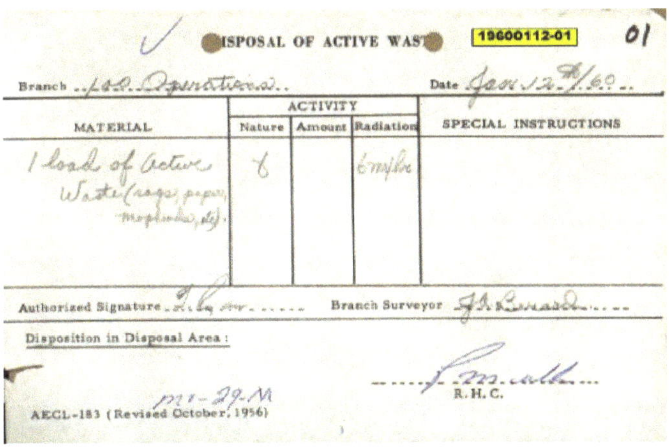
ISPOSAL OF ACTIVE WAST 19600112-01 01

Branch .... Date ....

MATERIAL	ACTIVITY			SPECIAL INSTRUCTIONS
	Nature	Amount	Radiation	
1 load of Active Waste (rags, paper, mop heads, etc).				

Authorized Signature .... Branch Surveyor ....

Disposition in Disposal Area :

.... R. H. C.

AECL-183 (Revised October, 1956)

FIG. 34. Waste Disposal Slip from the historical records at the Chalk River Laboratories site. Image courtesy of Atomic Energy of Canada Ltd.

of the content of the special disposals and any special instructions regarding that disposal.

- Historical photographs. The timing and location of specific burials can often be bracketed by examining time sequence photographs. This is particularly easy to do if aerial photographs are available. The Provincial Ministry of Natural Resources maintains a photograph library that contains aerial photographs from before the construction of the site. They have has also been routinely taking photographs since then. The Atomic Energy of Canada Ltd and Canadian Nuclear Laboratories have also been taking both aerial and ground photographs since the 1940s. These photographs have been useful in detecting ground disturbances in the outer wooded areas due to CRL activities, which may require remediation. They have also revealed the practice of using filled waste trenches as burning areas and have shown the location of asphalt trenches that are not marked on waste management area maps. The use of the earliest waste management area as a series of infiltration trenches for cooling water, following the NRX reactor accident in 1952, is also documented in historical photographs.
- Nuclear Material Accountability Forms. Fissile materials on the CRL site have always required strict tracking and accounting. The movement of these materials around the site and out to disposal areas is and has always been carefully documented. These forms can give an indication of how much fissile material is in a certain location.
- Maps and/or drawings. Drawings exist for the very first disposal into the ground at CRL in February 1946. The drawings are supposed to indicate the location of waste trenches and other special burials. The maps have been updated periodically over the years, but at irregular intervals, and changes to the waste management areas are known to have occurred without the drawings being updated. Engineering drawings also exist for engineered waste disposal structures such as ILW bunkers and tile holes.
- Operational records from waste treatment facilities. Several infiltration pits were used from approximately 1955 to 1995. Inventories of many contaminants can be estimated from records that operational production groups produced to record wastes from their processes.
- Technical and scientific reports. Over the years, staff have performed scientific experiments, pilot scale process mock-ups, run reactors, produced isotopes and processed irradiated fuel. As they engaged in these activities, they wrote reports to describe their work (the equipment and chemicals used) and the outcome of the activities. The waste generated from these activities ended up in waste trenches, liquid waste tanks and other 'special' burials. These reports are particularly valuable for the years prior to 1956, when disposal records were destroyed in a fire.

— Radiation protection surveyor logbooks. Radiation protection surveyor logbooks are prepared to document the safety of workers while undertaking radiologically hazardous work. These logbooks usually document radioisotopes, or at least gamma radiation fields, that may be present at a particular location, from which contamination levels can be inferred.
— Modern day groundwater monitoring reports provide information on what is escaping from the legacy trenches, indicating types of waste that were emplaced in the earlier operational phases of the trenches.

I.3.4. Uncertainty associated with historical records

Owing to the time period in question, many knowledge gaps and uncertainties exist. The state of the records, issues with using the records and associated uncertainty include the following:

— Non-existent records. Waste items were occasionally placed in pits or specially dug holes in the ground, with no records left to indicate their location. These items show up as anomalies when non-intrusive geophysical surveying is performed. Test pits to explore these anomalies often, but not always, reveal benign items with little or no associated hazard.
— Misleading records. In the past, different disposal facilities were called different names by different groups. Illegible handwriting can make disposal records hard to read. Locations of wastes, as shown on historical maps, are not always entirely accurate. Updates to drawings or maps were made infrequently or not at all.
— Incomplete records. Commonly, disposal slips recorded dose rates from the waste packages (e.g. 0.3 mSv/h) and did not provide any indication of the isotopes producing the radiation field. The waste description is often simply "truckload of waste". Contaminants may only be identified as "gross alpha" or "gamma" with no further identification. Information that was provided focused on that needed for the safety of operators emplacing the wastes or working nearby in the waste management areas. Recording information for future waste retrieval and disposal was not a concern at the time. Even if the burial of radiologically contaminated waste was recorded, most documentation contains little or no non-radiological, chemical information. That information was not recognized as valuable at the time. Neither the Waste Disposal Slips (Fig. 34) nor waste management area logbooks had a place to record the chemical information.
— Loss of records. Prior to 1956, waste management records were stored in a wood frame building (most buildings on-site were wood framed at the time). In 1956, a fire burned the entire building and the records that were

FIG. 35. Fire destroys the wood frame building containing waste disposal records in 1956 at the Chalk River Laboratories site. Image courtesy of Atomic Energy of Canada Ltd.

stored therein (Fig. 35). Still, there is evidence that records existed prior to 1956 because about half a dozen disposal slips have survived from that time period.

I.3.5. Managing and mitigating uncertainty

Methods to reduce the uncertainty inherently associated with older records from early operational time periods have been implemented at CRL. These methods include the following:

— Field sampling and analysis. Extensive characterization activities, both intrusive and non-intrusive, are carried out to get information for planning and decision making. Data collection includes radiological activity and concentration quantification. Locations of wastes in the field are also determined with greater accuracy than that provided in the waste records.

— Scientific records. A search of the technical and scientific reports from the 1940s and 1950s examined the site activities that were taking place at the time and hence inferred the wastes that were being sent for disposal. This search yielded both radiological and non-radiological information, as well as information on bulk waste forms, such as fume hoods and other large equipment (e.g. solvent extraction equipment, tanks).

— Using a CSM to indicate problem areas. The use of CSMs supports the development of the site end state and planning for remediation activities. If the location of contaminant sources is known, a CSM and the use of predictive models can help to determine the contaminant concentrations in places that are harder to access (e.g. wetlands). This information is subsequently used in the environmental risk assessment.

— Inference from records of other waste management areas on the site. The inventory for Waste Management Area A (the earliest waste management area, which operated from 1946 to 1954) can be inferred to a large extent from the inventory of Waste Management Area B (operated from 1953 to 1963) for those activities that spanned both time frames. In other words, the post-1956 records can be used to infer inventories prior to 1956, if the same processes were being performed before and after that date.

— Planning conservatively for remedial action implementation. Legacy records will never provide all the desired information to remediate a site. Risks associated with not having all the information have to be managed by having mitigating actions in place. Decisions to remediate may be optimized to account for information gaps. Mitigations to ensure the safety of field operations may include hold points under specified conditions, defence in depth tactics, etc.

— Fingerprinting[44] based on operational knowledge. Major contributors to waste in the past were the NRX and NRU reactors. Much of the waste on the site would have originated from one of the two reactors, and consequently they are expected to have similar ratios of contaminants. In some instances, hard to measure radioisotopes of significance may be inferred from reactor core inventory analyses (e.g. the activity ratio of ceasium-137 relative to zirconium-93 or antimony-126 or caesium-35).

I.3.6. Conclusions and lessons learned

For various reasons, the historical waste record at CRL is incomplete or inaccurate. Often, insufficient historical information is available to make informed environmental remediation decisions. Consequently, historical information often needs to be supplemented with data obtained from alternative sources (e.g. operational activities that produced records from which the wastes generated can be inferred, current field investigations).

The uncertainty associated with historical waste inventories can be managed using mechanisms such as the CSM and contaminant transport modelling and risk assessments. This approach is necessary because, as it is not possible to have all the required information, such mechanisms and the use of conservative decision making will allow progress towards site cleanup.

---

[44] Also known as scaling factors or vectors.

I.4. MEXICO, CENTRO DE ALMACENAMIENTO DE DESECHOS RADIACTIVOS

I.4.1. Site setting and description

Centro de Almacenamiento de Desechos Radiactivos (CADER) is the radioactive waste storage facility operated by the National Institute for Nuclear Research. The site receives LLW and ILW generated from medical, industrial and research activities.

The CADER site is located in the municipality of Temascalapa in Mexico, approximately 35 km from Mexico City. Several population centres are within a 1–8 km radius from the site.

The CADER site has a total area of 16.4 hectares and is divided into an uncontrolled area and a radiologically controlled area. The uncontrolled area comprises administrative and operational areas that are not related to the areas of radioactive waste storage. The controlled area contains five trenches (0, 1, 3, 5 and 7), 29 pits and three waste storage buildings (named I, II and III). Both the pits and the trenches are now closed. Figure 36 shows the site layout; the waste trenches are located in the north of the site and are separated by a field from the administrative area and the radioactive waste storage buildings.

Land use within a 10 km radius around CADER is mainly agricultural. The site is underlain by basalt spill, tuffs and lapilli rock. The geological and climatic conditions prevent the development of any significant surface water bodies, although there are streams and ponds. The underlying groundwater body is approximately 260 m below ground level. Groundwater is used for drinking water by the local population.

I.4.2. Operational history of the site

The CADER site began to operate as a disposal site in 1970. The disposal method was burial of radioactive waste (with or without conditioning) in trenches. Disposal operations ceased in 1989; since 1991, the site has operated as a waste storage facility.

The trenches were excavated directly into the natural soil and according to available information, the trenches do not have any engineering features. The exception to this rule is the concrete pits that were used to dispose of disused radioactive sealed sources. There are three concrete pits located in the footprint area of Trench 1 and Trench 26 with the footprint area of Trench 5. The approximate dimensions of the trenches vary from 174 m to 185 m in length, 1.5 m to 3.5 m in depth and are 1.3 m wide. The trenches are backfilled and covered by mounds and concrete caps to support water runoff with the soil excavated from

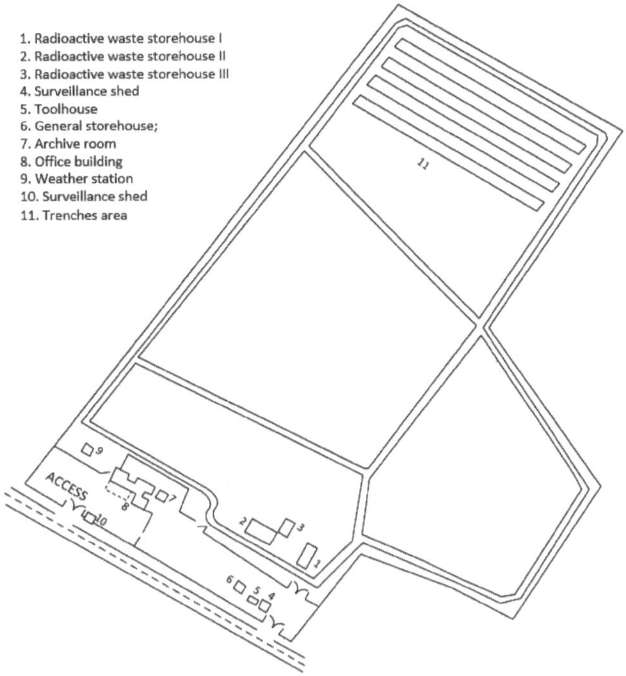

FIG. 36. Site layout of the Centro de Almacenamiento de Desechos Radiactivos. Figure courtesy of J. Anguiano (National Institute for Nuclear Research) and J. Martínez (Comisión Nacional de Seguridad Nuclear y Salvaguardias).

the trench. The ground around the trenches slopes to the north-west, and at the lower end of each trench there is a water collector for sampling purposes.

The types of radioactive waste that were buried in the trenches include uranium tailings, contaminated steel rods, compacted solids, biological materials, liquids, gelled liquids, waste sludges, contaminated soils, contaminated equipment and parts, contaminated resins, contaminated debris, activated components and spent sources (both immobilized and not immobilized).

I.4.3. Regulatory context

Secretaría de Energía de México (Ministry of Energy) is legally responsible for the storage, transport and disposal of nuclear fuel and radioactive waste. Secretaría de Energía de México is also responsible for approving the guidelines and programmes regarding nuclear industry activities, including those related to radioactive waste management. There have been several attempts to develop a national policy and strategy for the management of spent fuel and radioactive waste; an attempt was undertaken between 2012 and 2015 with support from the European Commission [184]. Without a national policy and strategy, neither the responsible organization nor the long term strategy for the site and the radioactive waste currently in storage has been defined.

At the time that the site commenced operations, Comisión Nacional de Seguridad Nuclear y Salvaguardias (CNSNS) had not been created, and there was no legal framework regarding radioactive waste management. According to available information, the first licence granted for the CADER site was for radioactive waste disposal. The licence was issued in October 1983 and the trenches stopped operating in 1989.

As a result of an extensive site study performed by the Electricity Federal Commission, it was concluded that the site was not suitable for a disposal facility owing to geological and demographic issues. As a consequence, in 1993 CNSNS changed the licence from a disposal site to a storage facility, where the burial of waste was explicitly prohibited. Since 1993, the site has been operating as a radioactive waste storage facility with a licence in force that has to be renewed every two years. The licence covers the operation of the three storage buildings and the surveillance of the closed trenches.

I.4.4. Drivers for site assessment

Since 1996, CNSNS has included a series of actions in the site licence related to better understanding of the status of the site. This has included the requirement to implement an annual environmental radiological monitoring programme.

I.4.5. Site evaluation

The main site evaluation works were undertaken in 2004. The starting point for the characterization of the trenches was an analysis of the existing inventory records, complemented with a sampling and measuring programme to validate the available information. The aim of the intrusive investigation was to obtain information on the waste condition and inventory and potential contaminant migration. This information could be used to understand whether the trenches posed a risk to people or the environmental and the technical requirements for waste recovery.

A sampling and analysis programme was conducted that required intrusion into the trenches in ten locations. Sampling was undertaken from different trench locations, with samples collected both near the surface and at the trench bases. Samples were collected from a range of different media, including soil, drums, concrete, contaminated materials and uranium tailings. The radionuclides of interest were potassium-40, cobalt-60, caesium-137, radium-226 and uranium-235.

The visual inspection of drums indicated that some drums were in good condition, whereas others were heavily damaged by corrosion. The damage to the drums was believed to be partly related to the poor quality of the drum material, the inclusion of liquids in the drums and water infiltrating into the trenches. All the drums observed at the lower level sampling points were extensively corroded.

As may be expected, high contaminant concentrations were recorded in waste samples. However, concentrations of contaminants measured in the underlying soil were relatively low [185]. The results of the assessment concluded that the trenches did not pose a significant impact on the environment. If the waste is retrieved in the future, localized soil contamination could be removed during remediation of the trenches.

The ongoing environmental radiological monitoring programme started in 1991 to monitor the trenches and the site's operational activities to ensure that there are no environmental impacts. The monitoring programme requires the collection of air, soil, superficial water, potable water, 'nopal' (comestible cactus) and sometimes bean samples. The samples are analysed for specific radionuclides, such as caesium-137 and tritium, as well as gross alpha and beta activity. CNSNS performs a verification environmental radiological monitoring programme, in parallel to that of the licensee. CNSNS's programme is aimed at comparing and verifying consistency between the licensee's and regulator's results.

I.4.6. Remediation options and selected approach

Until a national policy and strategy on radioactive waste management is developed and approved, a decision cannot be made on the long term future of the waste trenches. Furthermore, without appropriate government funding, a national waste repository or defined remedial criteria, it would be difficult to implement waste retrievals and remediation at the site. Since the trenches remain stable, retrieval and interim storage of the waste are likely to result in unnecessary radiation exposure to workers. This approach would also incur greater financial costs compared with directly disposing of the waste once a national disposal facility is available. The ongoing environmental monitoring programme is an essential requirement to ensure that the site remains stable and there is no environmental impact until the long term future of the site is decided.

I.4.7. Implementation of the remediation approach

The implementation of remediation has not yet been confirmed. Environmental monitoring and evaluation continue at the site as part of normal operations.

It is worth recognizing that any works undertaken at the site will involve the support of interested parties. Therefore, it is important to actively improve stakeholder confidence in the operation of the site during the current monitoring phase and in preparation for any future works. The previous experience gained during the remediation at the uranium mining and milling facility, Villa Aldama, proved that building stakeholder confidence was an essential determinant to the success of the project.

I.4.8. Conclusions and lessons learned

It is commonly said that "planning without acting is a dream but acting without planning is a nightmare". Mexico recognizes the need to establish a national policy and strategy for waste management and the associated decisions on funding and a national waste repository. Without these fundamentals, it is difficult to plan the long term management of radioactive waste, including that at the CADER site. Furthermore, whilst the trenches remain stable and there is no impact to the environment, it is difficult to justify waste retrieval and environmental remediation.

The situation at CADER illustrates that safety might not be the main driver for the management approach and that non-technical factors can highly influence the decisions regarding the future of a legacy trench site.

I.4.9. Acknowledgements

The following experts have worked on the preparation of this case study: R.A. Suárez Alvarado, D.A. Mut Chable, J. Martínez Lizardo. Y.C. Ortega Nava provided support to the research process, and Y. Alvarez Rico and R. Fabian Ortega provided a technical review of the document. All the persons mentioned are CNSNS staff.

I.5. UNITED KINGDOM, LOW LEVEL WASTE REPOSITORY, HISTORIC WASTE TRENCHES

I.5.1. Site setting and description

The LLWR is the UK's main facility for the disposal of solid LLW[45]. It is owned by the Nuclear Decommissioning Authority and the site operator is Low Level Waste Repository Ltd (LLW Repository Ltd). The repository comprises historic waste trenches, operational engineered waste vaults, a railhead, operational areas for waste conditioning, administrative facilities and currently unused areas of land, some of which are identified for potential future vault disposals. Some of the currently unused land contains structures and buildings from the site's former use as a Royal Ordnance Factory from 1941 to 1945.

The LLWR is an operational radioactive waste facility. This case study is focused on the historic waste trenches, which were filled more than 25 years ago.

The LLWR is located approximately 400 m from the coastline near the village of Drigg in west Cumbria. Figure 37 provides an aerial view of the site and highlights the key areas within and around the site. The coastline is adjacent to the site of special scientific interest, which is just off the edge of the photograph. The site occupies around 110 hectares and waste disposal operations take place in the 40 hectares at the north of the site.

The geology of the region around the site comprises Quaternary sediments that overlie Triassic Sherwood Sandstone bedrock. The area around the LLWR has been extensively investigated; approximately 650 site investigation boreholes have been drilled on and around the LLWR. The heterogeneous nature of the Quaternary sediments has significant implications on hydrogeology, contaminant transport and coastal erosion.

Extensive intrusive investigation has been carried out at the site over many years. This has led to the development of a detailed hydrogeological conceptual site model (see Fig. 38 for a schematic representation of the model). The ongoing monitoring programme continues to help to validate and refine the model.

A plume of tritiated groundwater contamination arising from the trenches is evident in both shallow and regional groundwater moving towards the coast. There is no water extraction between the site and the coast, and the discharges from the LLWR are considered to pose no significant risk to the environment or members of the public. The primary source of the tritium is the disposals of

---

[45] LLW is defined in the UK as: "Wastes having a radioactive content not exceeding 4 Gigabecquerels per tonne of alpha activity or 12 Gigabecquerels per tonne of beta/gamma activity" [186].

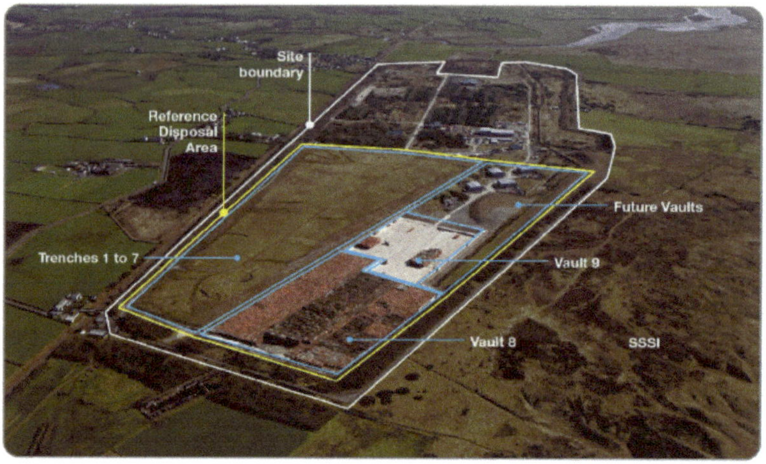

FIG. 37. Aerial view of the Low Level Waste Repository looking south, captured in March 2011. Photograph courtesy of Low Level Waste Repository Ltd. SSSI — site of special scientific interest.

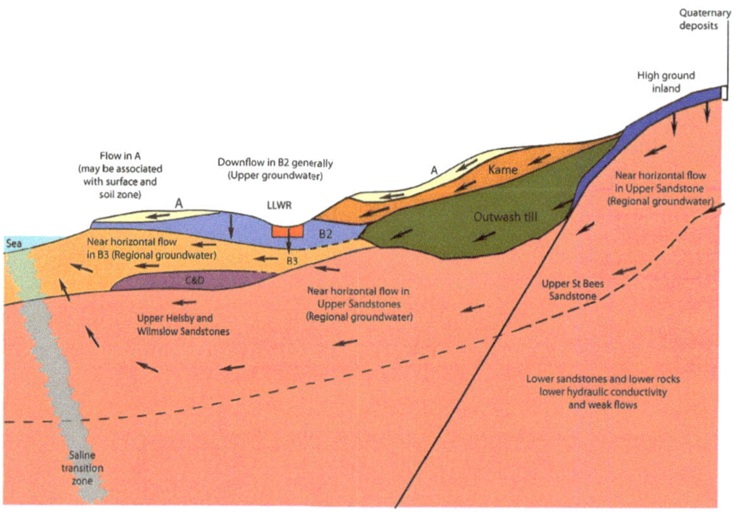

FIG. 38. The hydrogeological conceptual model of the Low Level Waste Repository, presented in the 2011 Environmental Safety Case. The schematic is the east to west cross-section and the arrows show the direction of groundwater flow. Figure courtesy of Low Level Waste Repository Ltd.

beta lights[46] in the trenches. The 2011 Environmental Safety Case (ESC) includes an assessment of the distribution and behaviour of the tritium plume to support understanding of the hydrogeology and contaminant flow [10]. The tritium plume was used as a marker to assist in the validation of groundwater models. In addition, the monitoring results have been used to estimate how much tritium has been released from the trenches.

A key focus for the safety assessment of the site is understanding coastal environment and processes, given the proximity of the site to the sea. Therefore, a substantial programme of scientific research and monitoring has been implemented to understand the current coastal system and to provide information for forecasting its future evolution. The results of both qualitative evidence and quantitative modelling studies have concluded that erosion of the vaults will begin on a timescale of a few hundred to a few thousand years and the subsequent erosion of the vaults and trenches will be complete within one to a few thousand years.

I.5.2. Operational history of the site

In the late 1950s, the site went into United Kingdom Atomic Energy Authority ownership as a nuclear establishment and was repurposed for radioactive waste disposals to support primarily the research activities on the Sellafield site. Disposals started in 1959 and solid LLW was tumble-tipped and buried in shallow trenches, similar to contemporaneous practice in the landfill industry (see Fig. 1).

Between 1959 and 1995, approximately 800 000 m^3 of waste was disposed of in seven trenches. Each trench was founded predominantly within an underlying clay layer that was intended to form a low hydraulic conductivity base. Bentonite was rotovated into the base of Trenches 5, 6 and 7 in areas where the clay layer was absent.

These trenches are now covered by an interim cap, which incorporates a plastic membrane to minimize water entry into the wastes. Leachate generated inside the trenches is currently collected along their bases. Leachate flows by gravity from the trenches to collection points and onwards to discharge to the sea via holding tanks.

The disposal of waste in containers placed in an engineered concrete vault began in 1988 and is ongoing in Vaults 8 and 9. A planning application has been accepted by the local authority and supported by the environmental regulators for the construction of Vaults 10 and 11, and the site is large enough to accommodate

[46] Beta lights were often used for exit lights and comprise gaseous tritium in a glass vial with a phosphor layer.

the construction of further waste disposal facilities. Continued operation of the site is expected until 2129 [187]. The site will be closed with the construction of a single, gently domed multi-barrier low permeability engineered cap, which will cover the trenches and the vaults. The final cap will be progressively constructed and designed for stability and resistance to erosion and will have low visual impact to fit into the local environmental setting. To minimize the visual impact of the LLWR, a long term vegetation cover will be created around and on top of the cap area.

The final status of the site will be a closed disposal site. The site will remain under institutional control, and it is assumed regulatory control, for a period of at least 100 years after the final disposals. From a sustainable development perspective, the reuse of the site is considered to be beneficial. Therefore, measures will be implemented to encourage appropriate interim uses of the site during the institutional control period. Uses are likely to include the increase of habitats to support biodiversity and limited recreational use. Longer term institutional controls may continue over the site in the form of covenants, planning controls and record keeping, in order to deter or prevent inappropriate site uses.

I.5.3. Regulatory context

The EA is responsible in England for regulating the disposal of LLW under the Environmental Permitting (England and Wales) Regulations 2016 (EPR16). The EA is responsible for permitting all landfill in England including LLWR and three landfills that accept very low level radioactive waste[47]. The Scottish Environment Protection Agency (SEPA) is responsible for authorizing disposals at the Dounreay Low Level Waste Vaults under the Radioactive Substance Act 1993 (RSA93).

The Office of Nuclear Regulation regulates LLWR regarding matters of nuclear safety. The site must comply with the requirements set out in the nuclear licence conditions. LLWR has a nuclear site licence, although other landfills accepting radioactive waste in the UK do not (e.g. the Augean facilities at East Northants and Port Clarence). The reason that the LLWR has a nuclear site licence is primarily a historic administrative process, rather than a technical need to operate under a nuclear licence. The process of surrendering a nuclear licence has not yet been tested for a radioactive waste facility and is not considered to be a priority activity whilst waste is being disposed of at the site.

[47] Very low level waste is defined in the UK as: "A sub-category of LLW, it comprises waste that can be safely disposed of with municipal, commercial or industrial waste or can be disposed of to specified landfill sites" [186].

As part of the permitting process, LLW Repository Ltd submitted an updated ESC in May 2011 to demonstrate that people and the environment are protected from the hazards associated with disposals to the facility now and into the future [10]. The ESC covers both the historic trenches and the operational waste vaults. The EA then completed a comprehensive technical review of the ESC to determine its adequacy as a submission against the permit requirements [188, 189] and whether it met the principles and requirements set out in the UK's Guidance on Requirements for Authorisation of Near-surface Disposal Facilities for Solid Radioactive Waste (GRA) [190]. In 2015, the EA issued a permit for continued disposal of LLW into the engineered waste vaults.

Under the requirements of the current environmental permit for the site, the EA expects the ESC to be maintained as a 'live' case, including annual periodic and major reviews. The EA also expects a forward programme of work to be developed with the aim of ensuring continued improvement to the ESC and continued compliance with the requirements of the GRA.

I.5.4. Drivers for site assessment

The key drivers for the ongoing assessment of the site are the requirements of the environmental permit and the nuclear site licence. The environmental permit requires environmental monitoring and the maintenance of the ESC, which includes closure of the trenches and the engineered waste vaults. LLW Repository Ltd carries out an extensive monitoring programme. The results of the programme are submitted to the EA annually [191].

I.5.5. Inventory

A key challenge faced in developing the ESC for the site was related to the uncertainties in the waste inventory in the historic waste trenches. The EA review of the 2002 ESC[48] concluded that better use should be made of the records available in order to derive the trench inventory and the trench inventory should not rely solely on the use of generic radionuclide fingerprints.

As part of the activities to improve the trench inventory, LLW Repository Ltd commissioned 'recall' interviews with past and present staff to identify any waste that may have been disposed of at the facility without accurate records to supplement the understanding of the inventory.

---

[48] The term 'post-closure safety case' was used prior to the adoption of the term 'Environmental Safety Case'.

To provide greater accuracy in the representation of the trench inventory, waste heterogeneity maps can be developed for key radionuclides and other important materials such as cellulose.

Deriving the inventory for the waste trenches is complex because of the variable extent and quality of the available information. The information in the historic records met the requirements at the time, but it is recognized that today's requirements need a greater level of detail. The approach taken has been to derive a trench inventory [192] that reflects both the quality of the available data and the actual or potential significance of individual radionuclides in the performance of the facility. To facilitate this approach, the available records were examined and divided into two categories: key consignments and routine consignments.

Disposals were identified as 'key consignments' if the quality of the information available was considered significant in terms on the inventory. The data used for the inventory for these consignments were taken directly from the disposal records. For these consignments, the quality of the information is believed to be comparatively high, as it is based on better characterization.

All other disposals, where the quality of the information was considered of lower quality than that for key consignments, were assigned as 'routine consignments' and were more commonly received at the site. Radionuclide fingerprints were derived from information gathered from the UK Radioactive Waste Inventory for characterized waste streams that were believed to be similar in nature. This allowed the disposal inventoried to be calculated by backfitting the information to the reported disposal volumes.

The consignment data have been used to evaluate the following:

— The expected material types and volumes per trench;
— The impact on human health and the environmental impact of relevant radionuclides according to the trench by trench inventories;
— The location and approximate concentration of important radionuclides;
— The location and volume distribution of key materials, including lead, rubble, ferrous metal, soil, asbestos, cellulosic materials and wood.

The waste heterogeneity maps were also used to develop potential remediation options using targeted retrieval of wastes (see Section I.5.7 for further information).

There was also evidence from the recall interviews that discrete items had been disposed of at the site that individually contained significant levels of radioactivity. It was possible to infer that these items were not widespread enough to significantly affect the assessment of impacts or results of the ESC.

Although uncertainties remain regarding the trench inventory, this uncertainty was accepted by the EA because of the nature of the historical disposal practices at the site. However, the EA expects the operator to make best use of any relevant sources of information that may become available, with the intention of reducing inventory uncertainty as far as is practicable.

I.5.6. Trench remediation options

One element of the ESC has been to evaluate whether waste retrieval and/or remediation measures are necessary to enable site closure. The options assessment considered a range of options and suboptions in the following groups of technologies: full waste retrieval; selective waste retrieval; in situ solidification and stabilization; ex situ solidification and stabilization; groundwater barriers; and an impermeable cap.

The retrieval of all wastes in the trenches was screened out because it was disproportionate in terms of cost, complexity, worker dose and environmental hazard compared with the long term risk posed by the waste to the public and the environment.

The conclusions of the inventory study and post-closure safety assessment indicate that the radionuclides that give rise to post-closure risks via gas and groundwater pathways are widely distributed throughout the trenches. The modelling indicated that the calculated impacts are related to the radionuclide average concentrations in large areas of the trenches. For the coastal erosion and human intrusion scenarios, the radionuclides present in relatively small volumes of certain wastes that are located in specific bays in the trenches dominate the post-closure impacts. These impacts dominate the average calculated doses and could potentially give rise to local impacts, above the average, across the site. Therefore, selective retrievals of these wastes were considered as a potential remedial option. Consequently, three retrieval options were formulated involving excavation of bulk waste from specific areas of the trenches. The specific areas were identified as areas where waste existed that could potentially negatively impact the post-closure safety case. It was assumed that, following sorting, the waste remaining would be retained and either returned to the trenches or conditioned and emplaced in the engineered vaults. Table 7 presents the three retrieval options and describes the potential improvements that they offer to reduce post-closure impacts.

The assumptions that underpinned the expected amelioration of post-closure impacts assumed that the retrievals would be one hundred per cent efficient in the removal and sorting of the target wastes. Removal efficiencies below this would provide amelioration of impacts on a pro-rata basis, but the actual retrieval efficiencies that might be achieved are very difficult to quantify. Therefore, the waste removal and sorting efficiency was an important source of uncertainty when deciding whether selective retrievals from the trenches are necessary.

TABLE 7. RETRIEVAL OPTIONS AND LEVELS OF AMELIORATION
(based on data from Ref. [193])

	Option 1	Option 2	Option 3
Description	Target waste retrieval of thorium sands (monazite and thorite) from two discrete locations in Trench 2.	Retrieval of radium-bearing process wastes from regions in Trenches 2 and 3.	Retrieval of thorium-bearing wastes more widely dispersed in Trenches 4 and 5.
Volume	7 000 m^3 excavated waste including 1 740 m^3 target waste.	36 800 m^3 excavated waste including 768 m^3 target waste.	82 200 m^3 excavated waste including 2 731 m^3 target waste.
Objective	Removal of highest concentrations of thorium-232, which present the highest localized dose rates from coastal erosion and intrusion (excavation) cases.	Removal of highest concentrations of radium-226, which present the highest localized doses from a possible intrusion (building on spoil) case.	Removal of more widely dispersed thorium-232, which, after remediation of Trenches 2 and 3, presents the next highest impacts.
Amelioration	Removes highest localized dose rates at the eroding cliff (c. 1 μSv/h). Recreational beach user case for Trench 2 is reduced from 0.011 to 0.001 mSv/a; the average over all trenches is reduced from 0.006 to 0.004 mSv/a.	Removes the highest doses for the most pessimistically located human intrusion building case (c. 20–30 mSv/a). Human intrusion building case assessed on a trench by trench basis reduced from 3 to 0.05 mSv/a for Trench 3 and from 1 to 0.25 mSv/a for Trench 2; average over all the trenches is reduced from 0.76 to 0.14 mSv/a.	Reduces dose along both Trenches 4 and 5 from 0.009 to 0.004 mSv/a; average over all the trenches is reduced from 0.006 to 0.004 mSv/a. Assuming options 1 and 3 are enacted, the average dose from recreational beach use over all trenches is reduced from 0.006 to 0.002 mSv/a. For all three options enacted, it is reduced to 0.001 mSv/a.

TABLE 7. RETRIEVAL OPTIONS AND LEVELS OF AMELIORATION *(based on data from Ref. [193]) (cont.)*

	Option 1	Option 2	Option 3
Results	Factor of ten reduction in maximum annual dose. Reduction of assessed risk of 30%. However, the assessed risk is already below the annual risk guidance level[a].	Factor of about ten reduction in maximum intrusion building case dose. However, the maximum assessed dose is already at the annual dose guidance level[b].	Assuming options 1, 2 and 3 are enacted, then there is a reduction of assessed risk of ~60%. However, the assessed risk is already below the annual risk guidance level[a].

[a] Annual risk guidance level of 10^{-6}, which is equivalent to a dose of 0.02 mSv/a (see Ref. [190] for further details).

[b] Annual dose guidance level of 3 mSv/a for prolonged exposure (see Ref. [190] for further details).

Options 1, 2 and 3 would reduce doses and assessed risk; however, it is important to note that the assessed doses and risk levels were already below those required by the GRA [190]. This output was then included into the optimization process where the options were further assessed against a broad range of factors including worker dose, conventional health and safety risks, financial cost and environmental impact. The options assessments concluded that the feasible scale of dose or risk reduction, below what are already low doses or risks by comparison with the regulatory requirements, was small compared with detriments of implementing the retrievals.

I.5.7. Development of remediation options

The focus on the development of remediation options at the site is to maintain safety during the continued site operations and to eventually enable site closure.

Since 2010, LLW Repository Ltd has improved the perimeter drainage infrastructure of the trenches and the measurement capability of the trench probes. After collecting three years of data, the trench water balance model was updated to provide better quantification of the volume of rainfall entering the trench disposals. This work showed that the interim trench cap was not performing as well as assumed in the 2011 ESC. Subsequent intrusive investigation work showed that this was primarily due to damage to the membrane during construction of the interim cap. This subsequently led to a programme of work to assess the need to improve the interim trench cap before the placement of a final cap. The calculation of the trench water balance can also help to validate the performance of any other remedial measures prior to the installation of the final cap.

In order to continue to demonstrate optimization, consideration of alternative management approaches will continue as new information or new technologies become available. However, the current optimal approach is to construct an engineered final cap and cut-off walls and repair the damaged interim trench cap membrane. This could deliver improved long term environmental protection. The function of the cap is to limit infiltration into the waste and thereby reduce the volume of leachate generated in the trenches; to provide a barrier to reduce radon release; and to restrict, in the long term, potential for intrusion into the trenches. Figure 39 shows a schematic of the key engineering features for the site closure. This final capping process has received planning permission. The project to emplace a final cap is part of the Repository Development Programme and in total it is expected to cost approximately £86 million. The works will be split into twelve phases over the coming decades.

Phase 1 of the Repository Development Programme, comprising the emplacement of the first strip of the final cap to cover the northern area of the trenches and Vault 8, received financial approval from the Nuclear Decommissioning Authority in August 2016. It is anticipated that construction of the first strip will be completed in 2028. Meanwhile, work will also be undertaken to repair areas of the current interim cap over the trenches. It is anticipated that

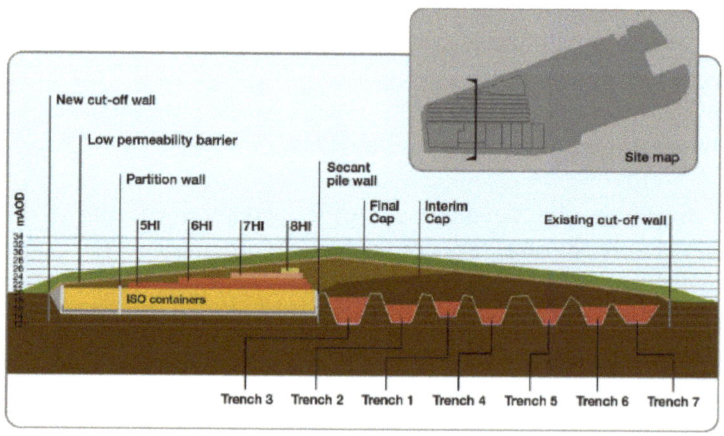

FIG. 39. Schematic of the key engineering features for the closure of the Low Level Waste Repository, UK. Figure courtesy of Low Level Waste Repository Ltd.

repairs to the interim cap on the trenches will be completed over a similar time frame. Overall, phase 1 comprises the following works:

- Perform enabling works (including tree clearance, screening for local stakeholders, construction of the cap shoulder, importation of materials, preparatory works for a large construction project lasting 7–10 years);
- Surcharge the north end of the trenches;
- Increase the stack height of waste containers in Vault 8 using waste containers from Vault 9;
- Complete the cap profile;
- Install the first strip of the final engineered cap;
- Repair the rest of the interim trench cap.

During the work, optimization and engineering assessments will be ongoing to address the monitoring approach before, during and after construction of the first stage of the final cap.

I.5.8. Conclusions and lessons learned

There are several lessons learned and conclusions drawn for the management of the historic trenches in this case study:

- Best use should be made of any relevant sources of information that may become available, with the intention of reducing inventory uncertainty as far as is practicable.
- Anecdotal evidence obtained from former site employees can be a valuable source of information on the past practices and inventory.
- The inventory studies provided information that allowed decisions to be made on whether trench remediation was necessary. The scale of dose or risk reduction in the long term needs to be compared against a broad range of factors, including worker dose, conventional health and safety risks, financial cost and environmental impact in the short term.
- Ongoing assessment of the site is a key requirement of the environmental permit and the nuclear site licence. This may require remediation options to be implemented for non-operational areas of the site.

Environmental monitoring can be used to assess the performance of the site. For the LLWR, it has been identified that to deliver long term environmental protection, an extensive programme of capping is needed. This will take several years to be delivered but is expected to limit infiltration into the wastes, reduce the impacts of radioactive gas release and, in the longer term, limit the potential for intrusion into the wastes.

I.6. UKRAINE, RED FOREST RADIOACTIVE WASTE TEMPORARY STORAGE SITE

I.6.1. Site setting and description

The Red Forest (Ryzhi Les) Radioactive Waste Temporary Storage Site (RWTSS) was created in the aftermath of the accident at the ChNPP, primarily to dispose of contaminated trees and soils from the surrounding forest.

The State Specialized Enterprise ‘Central Enterprise on Radioactive Waste Management’ (CERWM) is currently responsible for site maintenance and remedial measures. The State Agency of Ukraine on Exclusion Zone Management is responsible for managing the Chornobyl Exclusion Zone. Safe management of radioactive waste storage sites, such as the Red Forest RWTSS, situated within the Chornobyl Exclusion Zone, are within the overall responsibility of the State Agency of Ukraine on Exclusion Zone Management.

The Chornobyl Exclusion Zone is a restricted area comprising an inner 10 km zone and the remaining 30 km zone. It is assumed that within the 10 km zone, controls will remain in place for several hundred years to enable the decommissioning of Unit 4 of the ChNPP to achieve radiologically safe conditions.

The Red Forest RWTSS is situated in the 10 km zone and is approximately 1.5 km from the plant. The site is situated within the highly contaminated ‘western trace’ of the radioactive fallout, which was formed immediately following the explosion of Unit 4 on 26 April 1986. A specific feature of this trace is the presence of a large amount of micrometre sized ‘hot particles’ formed by the destruction of the nuclear fuel in the explosion of the ChNPP Unit 4 reactor. The important radionuclides associated with fuel particles include caesium-137, strontium-90, plutonium isotopes and americium isotopes.

At the time of the accident, the territory adjacent to the ChNPP was covered by 30–40 year old pine forests. As a consequence of the accident, the forest trees received a radiation dose of up to 10–100 Gy, which is a lethal dose for pine trees. The trees turned red-brown, and hence the forest has been known as the ‘Red Forest’ since then. Between 1987 and 1988, the civil defence troops carried out the in situ burial of the Red Forest. The primary objective of these site cleanup measures was to reduce external dose rates for workers involved in the construction of the sarcophagus and other works at the ChNPP site, as well as to mitigate the risk of wildfire in the dead pine forest.

The types of waste buried at the site include radioactive contaminated trees, topsoil, forest litter and some building debris (e.g. residential houses known as dacha), which were bulldozed into the trenches or into mounds. Upon completion of the cleanup operations, the territory with the waste burials was covered by a 30–50 cm thick 'clean' soil cover layer. The cover layer was formed from the sandy soil excavated from the trenches during their construction, as well as soil imported from nearby sand pits.

In 1989, forest planting (pines, birch trees) was carried out at the waste burial site to prevent wind resuspension and water erosion of the soil cover layer. As a result, the site is currently covered with an approximately 26 year old pine forest with mixed birch trees and bushes.

The radioactive cleanup work between 1987 and 1988 was not properly documented. The location of the trenches was not marked out on the ground and the radioactivity inventory was not accurately measured. Since 1991, characterization works at RWTSS have been undertaken to map the individual waste burials and to estimate the radioactivity inventory of the buried wastes.

The Red Forest RWTSS is situated in the central part of the first terrace of the Pripyat River (Fig. 40). The ground surface is largely flat, varying in height from 112 m to 115 m above the level of the Baltic Sea. The site is underlain in many places by human made deposits, such as soils and construction debris replaced during the cleanup operations. This is underlain by sandy Quaternary (Upper Pleistocene and Holocene) deposits to depths of 30 m below ground level, which are subdivided into alluvial and aeolian suites of deposits. Underneath

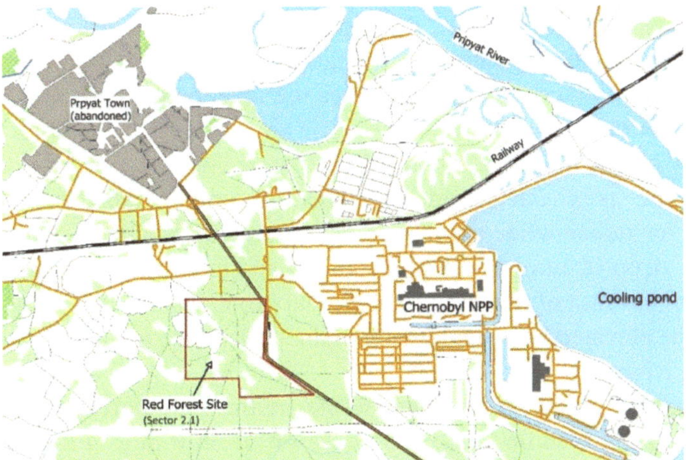

FIG. 40. Location of the Red Forest Radioactive Waste Temporary Storage Site. Figure courtesy of D. Bugai, Institute of Geological Sciences, Kiev.

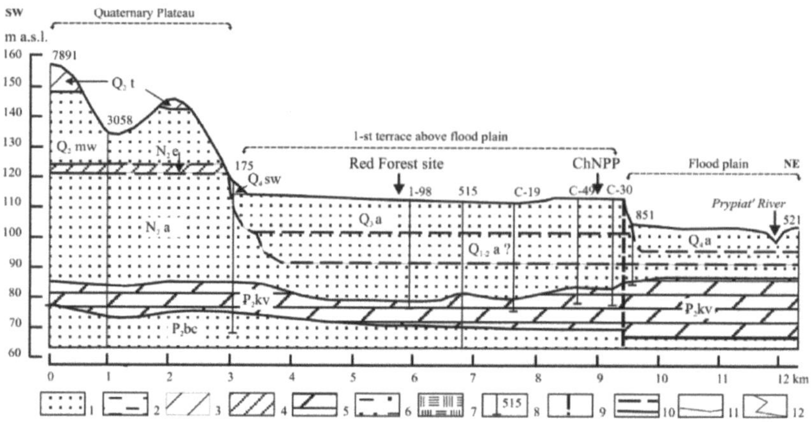

FIG. 41. Regional geological cross-section of the Chornobyl nuclear power plant (ChNPP) territory (reproduced with permission from Ref. [194]). Key: 1 — sands; 2 — silts; 3 — basal till; 4 — clay; 5 — marl; 6 — inter-bedding of sands and silts; 7 — peat and peaty sand; 8 — boreholes (numbered); 9 — inferred fault; 10 — boundaries between suites; supposed (upper) and established (lower); 11 — boundaries between depositional facies; 12 — facial replacement; 13 — groundwater level (generalized). Indices: Q_4 — Holocene; Q_{3-4} — Upper Pleistocene–Holocene unstratified; Q_3 — Upper Pleistocene; Q_{1-2} — Lower Pleistocene–Middle Pleistocene unstratified; N_2 — Pliocene; P_2 — Eocene; kv — Kyiv; bc — Buchack. Genetic types of deposits: a — alluvial; mw — meltwater; eol — aeolian; e — presumably waste mantle; sw — slopewash. Facies: ob — overbank; ch — channel; a-ch — abandoned channel.

these, lies the regional aquitard layer composed of marls (carbonate clays) of the Kiev suite of the Eocene epoch. Figure 41 shows the regional geological cross-section of the territory and indicates the location of the Red Forest and of the ChNPP. Further detailed information on the geology and geomorphology of the territory can be found in Ref. [194].

The site is characterized by generally unfavourable hydrogeological conditions for radionuclide retention. The groundwater table is shallow, and the local sandy deposits have a high permeability and low sorption capacity. The site is located in the area of transit of the regional groundwater flow system from the region of the elevated Chistogalovka moraine hills to the main discharge point of the Pripyat River.

The depth of the groundwater table in the Red Forest RWTSS varies, depending on the specific location and season, from 0 m to 4 m below ground level. In the north-west part of the area, seasonal wetlands can be observed, for example, after a snowmelt in spring (see Fig. 3).

I.6.2. Operational history of the site

An area of approximately 12 km^2 known as the 'zone of emergency cleanup' was subject to cleanup activities following the accident. The Red Forest, known as Sector 2.1, covers approximately 8% of the emergency cleanup area. The Red Forest RWTSS was created between 1987 and 1988.

A total of 61 waste burials have been detected in Sector 2.1 comprising 8 clamps[49] and 53 trenches. The trenches were excavated directly into the local sandy soil. The length of waste burials ranges between 20 m and 410 m, with a typical length of 100–200 m. The width of the burials varies between 8 m and 12 m, and depths are between 1.5 m and 3 m. The height of the clamps ranges between 1.3 m and 2.6 m [195].

In addition to the radionuclides present in the waste trenches and clamps, radioactively contaminated material is contained in the topsoil and subsoil (0.05 m to 0.5 m) layers. The residual surface contamination or hot spots are the result of the emergency cleanup activities in 1987 to 1988, which were carried out quickly and in difficult conditions.

I.6.3. Regulatory context

The regulatory agency is the State Nuclear Regulatory Inspectorate of Ukraine. Red Forest RWTSS is a licensed site for temporary storage of radioactive wastes, along with several other similar RWTSS sites within the Chornobyl Exclusion Zone.

The disposal facilities and the RWTSS within the Chornobyl Exclusion Zone are managed and maintained by the CERWM. The licence requires the CERWM to undertake monitoring and maintenance works, as well as to identify and implement measures to improve the radiation safety of the waste storage sites. Such measures may include retrieval and redisposal of wastes to engineered facilities. Any proposed remediation activity at the site is documented as a 'Technical Decision'. The Technical Decision is developed in consultation with stakeholders, which may include peer review support from the IAEA, and is approved by the regulatory authorities.

I.6.4. Drivers for site assessment

The site licence requires ongoing monitoring, assessment and, where possible, safety improvement of the site.

[49] A 'clamp' is the term used in Ukraine to describe an above ground mound of waste material.

I.6.5. Site evaluation

The following sections provide an overview of the site investigation and assessment works completed for the Red Forest RWTSS.

I.6.5.1. Previous characterization projects

The Red Forest RWTSS was surveyed from 1991 to 1992 by the VNIPIPT Institute (Moscow) [195]. During this survey, individual waste burials were mapped (Fig. 42), and their volume and inventory of radionuclides were estimated. Hydrogeology conditions and groundwater contamination by radionuclides in the vicinity of waste burials were studied.

Since 1994, the Institute of Geological Sciences (Kiev) has carried out groundwater monitoring studies at the Red Forest RWTSS as part of radioecology studies funded by the Ministry of Chornobyl Affairs of the Ukraine.

Between 1999 and 2012, comprehensive radioecological studies were undertaken, which included assessment of the type and extent of groundwater contamination, radionuclide speciation inside the waste trench and radioactivity accumulation in vegetation. The studies included many experiments conducted at an instrumentation laboratory installed at trench number 22 (T-22) of the Red

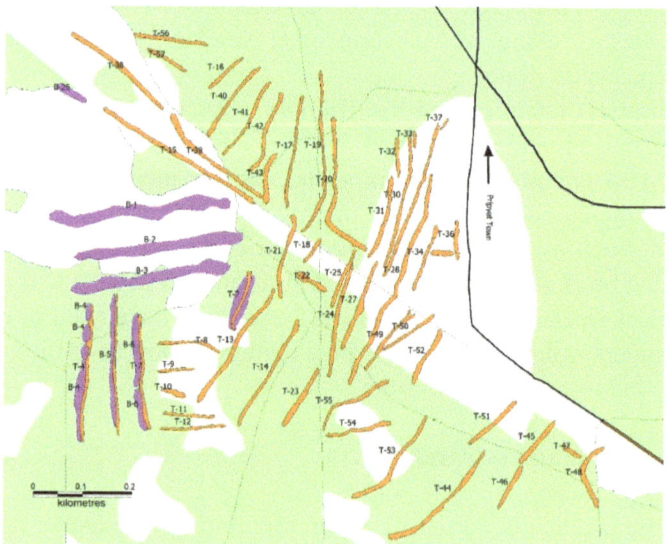

FIG. 42. Location of the waste burials in the Red Forest Sector 2.1 Radioactive Waste Temporary Storage Site. 'T' denotes a trench and 'B' a clamp (based on data of Ref. [195]). Figure courtesy of D. Bugai, Institute of Geological Sciences, Kiev.

Forest RWTSS. The work was carried out by a team of international radioecology experts under the Chornobyl Pilot Site Project (1999–2004) and the Experimental Platform in Chornobyl project (2005–2012) [93, 196–200]. The projects were funded by the Institute for Radioprotection and Nuclear Safety, which is a technical research organization sponsored by the French government.

I.6.5.2. Trench characterizations

The Red Forest RWTSS was surveyed in 1991 to 1992 by the VNIPIPT Institute (Moscow) [195]. The field procedures used by the VNIPIPT Institute to locate and investigate the waste burials included the following:

- — Visual inspections, for example, noting soil subsidence and vegetation anomalies;
- — Using georadar equipment;
- — Drilling of 1 m deep survey boreholes on a regular grid along the 40 m spaced profiles;
- — Measurement of gamma dose rate inside the survey boreholes.

From the gamma dose rate measurements, subsurface regions containing radioactive waste materials were identified[50] and their volumes were estimated. The caesium-137 activity concentration of the waste material was calculated from the gamma dose rate measurements using an empirical formula. Empirical correlation coefficients[51] were then used to calculate the activity concentrations of other radionuclides, such as strontium-90, plutonium-239 and plutonium-240, present in the fuel-containing waste.

On the basis of the results of the characterization studies, the median value of the total gamma and beta activity of waste materials (e.g. trees, soil) in burials in the Red Forest RWTSS was 1.6 MBq/kg, calculated as of 2000. The main radionuclides that contributed to the activity were caesium-137, strontium-90 and plutonium-241. The median value of the total alpha activity of waste materials was 80 kBq/kg and the main contributing radionuclides were plutonium-238, plutonium-239, plutonium-240 and americium-241.

The total amount of waste in the trenches and clamps in the Red Forest RWTSS was estimated at 150 000 m^3. In addition, about 50 000 m^3 of waste was contained in the contaminated topsoil layer hot spots. The total amount of waste

---

[50] Material was classified as radioactive waste if the gamma dose rate exceeded the relevant activity criteria for LLW.

[51] An approach similar to the use of scaling factors (also known as fingerprints or vectors) for waste characterization.

material in all RWTSS sectors encompassing the whole area of post-accident cleanup is estimated at 1.5 million m^3 [200].

I.6.5.3. Hydrogeological studies

The studies of the VNIPIPT Institute [195] established that 37 out of 49 trenches and the lower sections of 2 out of the 8 clamps were seasonally flooded to some extent by groundwater. In some of the trenches, the groundwater table was 1 m to 1.5 m above the base of the waste burials. These hydrogeological conditions support the leaching and migration of radionuclides from waste burials to the groundwater system.

Detailed hydrogeological studies of the Red Forest RWTSS site were carried out at an instrumentation laboratory installed at trench 22-T [93, 200]. These studies included characterization of the structure and hydraulic properties of the near surface sediments, groundwater level regime and regime of moisture flow in the unsaturated zone.

I.6.5.4. Groundwater transport of radionuclides

Comprehensive monitoring studies of groundwater contamination were carried out between 1999 and 2012 at an experimental site at trench 22-T. Data from trench 22-T indicated high mobility of strontium-90 in groundwater, with concentrations of approximately 1000*n*–10 000*n* Bq/L, relatively low mobility of caesium-137 with concentrations of approximately 0.01*n*–0.1*n* Bq/L, and low mobility of plutonium-239 and plutonium-240, with concentrations of approximately 0.001*n* Bq/L; where *n* denotes a number in the range from 1 to 10 [93, 199]. The strontium-90 plume in the aquifer, with concentrations of approximately 1000–2000 Bq/L, extends approximately 10 m downstream from the trench. Further information on the migration of strontium-90 from the trenches can be found in Ref. [199].

I.6.5.5. Radionuclide transfer to vegetation

Besides the hydrogeological migration of the radionuclides, studies also recorded the biogenic migration and accumulation of strontium-90 and caesium-137 in the biomass of the plants growing at the burial sites. The average pine tree plantation density for the Red Forest RWTSS area was estimated at 3300 trees per hectare [201]. The radionuclide biomigration processes from waste burials were studied in detail between 2000 and 2005 at the experimental site situated in the vicinity of trench 22-T [196, 201, 202]. Biogenic migration causes significant fluxes of radionuclides from the trench to the ground surface.

I.6.6. Overview of environmental modelling risk assessment studies

The risk assessment and modelling history of the Red Forest RWTTS encompasses nearly a 30 year period. Initial risk assessment analyses were focused on the groundwater pathway from the radioactive waste dumps and the related off-site risks caused by radionuclide transport to the Pripyat–Dnieper River system [195, 203, 204]. These analyses have shown that off-site risks caused by the groundwater pathway from strontium-90, which is the main contaminant of concern for this pathway, are generally low. This is a combined result of strontium-90 decay, the long radionuclide travel time in the subsurface, and the dilution of radioactive discharges in the Pripyat River surface water system.

The understanding of the source term model and relevant exposure pathways was significantly enhanced between 2000 and 2009 by research projects funded by the Institute for Radioprotection and Nuclear Safety. In particular, bioaccumulation of radionuclides (mainly strontium-90 and caesium-137) growing on the top of the trench was identified as an important radionuclide redistribution process in the waste site system [196, 202]. Studies carried out between 2005 and 2009 resulted in the development of a more sophisticated conceptual source term model, which accounted for geochemical evolution of organic waste material in the trench and for biocycling of radionuclides in the waste site system [93, 196].

Latterly, the focus of modelling and risk assessment analyses has shifted to consideration of the biouptake pathway from the trenches, atmospheric transport of radioactivity (forest fire scenario), recycling of activity to the topsoil with biouptake, and external exposure from the contaminated vegetation [94, 205].

A recently completed European Commission funded technical assistance project focused on risk assessment and the development of a risk based radioactive waste management strategy for the Chornobyl Exclusion Zone RWTSS (see Refs [94, 205]). The project was carried out by a consortium of European Commission consultants, while the beneficiary and end user of the project were the State Agency of Ukraine on Exclusion Zone Management and the Central Enterprise on Radioactive Waste Management. The developed remediation strategy for RWTSS was coordinated and approved by the State Nuclear Regulatory Inspectorate of Ukraine.

The safety assessment calculations were performed for a set of reference persons, including an unprotected on-site worker, a settler outside the 30 km zone, a settler inside the 30 km zone (e.g. Chornobyl town resident) and an inadvertent self-settler (e.g. settler in the RWTSS area). The risk assessment considered all relevant pathways (e.g. groundwater pathway, atmospheric pathways, external exposure, internal exposure due to consumption of contaminated foodstuffs). The scenarios analysed included both normal evolution and accidental scenarios (e.g. forest fire, tornado, intrusion scenarios). The time frame of assessment was 1000 years or until the maximum dose is reached through the relevant pathway. It was assumed that for the 10 km zone of the ChNPP, the institutional control (e.g. restricted public access) would be maintained for the next 500 years. This assumes 200 years of decommissioning and removal of radioactive debris followed by 300 years of institutional control of near surface disposal facilities.

I.6.7. Remediation options and selected approach

In the years following the accident at the ChNPP, the remediation approach proposed for the RWTSS was to retrieve all waste burials and collect the contaminated topsoil in the emergency cleanup zone. These wastes would then be disposed of at purpose built facilities at the 'Vector' complex in the Chornobyl Exclusion Zone. The volume of this material was estimated at 1.5 million m^3. This approach was based on formal consideration of the RWTSS inventory as 'radioactive waste' (according to activity concentration criteria) with all applicable radiation safety requirements.

However, recent studies have concluded that, from the perspective of safety for workers and the population, retrieval and redisposal is not a practical solution. The risk assessment analyses have led to the following conclusions [94, 205, 206]:

- For the majority of waste burials, robust institutional control (e.g. restricted public access) within the boundaries of the current 10 km radius zone of the Chornobyl Exclusion Zone over a period of about 500 years would be sufficient to ensure protection of the population, workers and the environment from the radiological impacts of the RWTSS. After this assumed institutional control period, the estimated residual radiological risks are sufficiently small, so that most land use restrictions may be lifted. In particular, the non-nuclear industrial use of RWTSS sites by unprotected workers would be possible (e.g. the estimated doses are less than 1 mSv/a).
- The retrieval of a few selected waste burials and/or topsoil contamination hot spots, including those situated within the Red Forest RWTSS, may be justified in some specific cases. For example, to improve workers' and visitors' safety, where the waste burials are situated in areas of prospective construction works and/or frequented by visitors. The need to remediate such individual waste burials will be evaluated on a case by case basis, in accordance with an overall dose optimization principle.
- Some waste burials containing higher level radioactive materials may still represent a residual risk (more than 1 mSv/a) for some scenarios (e.g. residential scenario) after the assumed institutional control period.

These residual risks can be potentially mitigated by retrieval of the waste material. However, as long as there is no high frequency presence of staff or visitors in the immediate vicinity of such waste burials, there is no need for any early waste retrieval operations. The benefits of deferring waste removal can be achieved because of the continuing decay of the activity inventory of these waste burials.

I.6.8. Implementation of the remediation approach

The approach to managing RWTSS in the Chornobyl Exclusion Zone described in Section I.6.7 was approved by regulatory authorities, and it is currently being implemented by the CERWM.

I.6.9. Conclusions and lessons learned

The presented case study can be instructive in several aspects. The poor knowledge management practices in the early phase of the creation of the RWTSS led to the loss of information on the location, design and inventory of waste burials. This resulted in the need to subsequently conduct costly characterization works, both in terms of financial cost and health and safety risks to staff, that have spanned several decades.

The remedial measures conducted at RWTSS at different periods resulted in several unforeseen negative side effects. The cleanup at the ChNPP site in 1987–1988 reached its primary objective of reducing the external dose rates by a factor of about ten. However, the creation of subsurface waste trenches resulted in serious groundwater contamination issues in the subsequent years. Similarly, while planting of the forest on top of the trenches, carried out between 1988 and 1989, reached its primary objective of stabilizing soil cover and minimizing dust resuspension, the side-effect was an intensive radionuclide transfer to vegetation from the buried contamination. This shows the importance of basing remedial measures on a comprehensive assessment accounting for all potential migration and exposure pathways and for possible interplay between these factors.

The Red Forest RWTSS case study shows that a risk based approach to planning remedial measures can lead to drastic change of the site management strategy, significant optimization of remedial efforts, reduction of waste volumes and a decrease in the use of other resources.

I.7. UNITES STATES OF AMERICA, MAXEY FLATS NUCLEAR DISPOSAL SUPERFUND SITE

I.7.1. Site setting and description

The Maxey Flats Disposal Site[52] (hereafter Maxey Flats) is a closed LLW disposal site owned by the Commonwealth of Kentucky. The site was licensed to the Nuclear Engineering Company Inc. (NECO) in 1963 for the disposal of LLW. The site was officially closed in 2018 after completion of remediation works, including the installation of an engineered cap.

Maxey Flats is located in Morehead, Kentucky. The overall site occupies approximately 310 hectares of land, which includes a buffer zone of 93 hectares. Approximately 136 000 m^3 of LLW was buried in an area of approximately 18 hectares, which is designated as a restricted area. Within the restricted area, approximately 11 hectares were used for the construction of the disposal trenches. The restricted area also contains storage and warehouse buildings, liquid storage tank buildings, gravel driveways and a parking area.

The local area is characterized by a series of hills and flat topped ridges and the disposal site is located on a spur of one of the ridges. The site is bordered to the east, south and west by steep slopes descending approximately 100 m to the valley floor. The land around the site is sparsely populated and primarily comprises woodland and open farmland.

The underlying geology comprises predominantly interbedded sandstones and mudstones.

I.7.2. Operational history of the site

Maxey Flats operated between 1963 and 1977 and approximately 127 000 m^3 of waste was disposed of at the site. The site accepted waste from a variety of organizations, including industrial sites, hospitals, research institutions and laboratories, as well as Department of Defense and US DOE sites.

The waste was disposed of in 46 unlined trenches, which were up to 207 m long, 21 m wide and 9 m deep. The trenches were used to dispose of both solid and solidified liquid wastes. Solid wastes included wooden crates, concrete boxes, metal drums and loose wastes, which were tumble-tipped into the trenches. The trenches were covered with 1–3 m of soil.

A series of structures known as ‘hot wells’ were also constructed for the disposal of small volumes of higher activity waste. These structures comprised a 3–4.5 m concrete coated steel pipe and were capped with a concrete slab.

---

[52] https://cumulis.epa.gov/supercpad/cursites/csitinfo.cfm?id=0402139

The disposed waste material was quite variable and included material contaminated as a result of neutron activation, approximately 250 000 kg of source material, 430 kg of special nuclear material[53] and at least 63 kg of plutonium.

Six further trenches were excavated after 1977 to receive waste material generated on-site as a result of the waste disposal operations.

An environmental monitoring programme was started in the early 1970s. The results of the monitoring indicated that radionuclides were leaching from the waste and migrating through the shallow groundwater. This information supported the commonwealth's decision to direct the Nuclear Engineering Company Inc. to stop operations in 1977. The Nuclear Engineering Company Inc. was responsible for the disposal licence and remediation of the Maxey Flats site. The disposal licence was terminated in 1979 and responsibility for the site was returned to the commonwealth.

Between 1979 and 1986, private organizations were hired to stabilize and maintain the site. This included environmental monitoring activities, the installation of a temporary cover that spanned the 11 hectare trench area and the installation of surface water controls.

In 1986, the site was placed on the National Priorities List[54] by the EPA. After a five year period of stakeholder consultation, site evaluation and planning, the EPA issued the Record of Decision, which set out the remedial plan for the site to enable site closure.

The physical remedial works were completed in 2016 and the EPA Certification of Completion was received in 2018.

I.7.3. Regulatory context

Prior to the inclusion of the site in the National Priorities List, regulation of the disposal activities was the responsibility of the commonwealth. As a National Priorities List site, the EPA took responsibility for ensuring that the site met the national regulatory requirements under the Comprehensive Environmental Response, Compensation, and Liability Act (CERCLA), whilst the commonwealth is the tri-party regulator and long term site steward.

Since the completion of remediation, four five-year reviews [207] have been conducted to confirm that remedial measures are still performing as expected and the site is being maintained appropriately. The project records from the CERCLA intervention are managed by the US DOE Office of Legacy Management.

---

[53] Plutonium, uranium-233, enriched uranium and uranium-235.

[54] The National Priorities List is managed by the EPA and comprises hazardous waste sites that are to be dealt with under the Comprehensive Environmental Response, Compensation, and Liability Act (CERCLA; also known as Superfund).

I.7.4. Drivers for site assessment

The key driver for assessing the site was the results of the environmental monitoring programme, which identified contaminant migration through leaching and movement through the shallow groundwater body.

In 1987, an Administrative Order by Consent was signed by 82 parties to initiate an impact assessment and remedial feasibility study for the site.

I.7.5. Site evaluation

Groundwater monitoring and surface water sampling has consistently been undertaken since the early 1970s. The site monitoring and development of the hydrogeological model indicated that the emergence of leachate from the trenches is via the bath-tubbing effect.

Additional monitoring and site investigation were undertaken in the late 1980s to underpin the Record of Decision[55] [5], which sets out the remedial plan for the site. Twelve radionuclides and eleven non-radioactive contaminants were identified from the soil, groundwater and surface water samples collected from the site. Tritium was the most abundant and mobile of the radionuclides and was identified as the primary contaminant of concern.

Environmental modelling and risk assessments were also conducted to underpin the Record of Decision. Following the development of the site conceptual model, modelling software such as MODFLOW and PATHRAE were used to support the dose and risk assessment. The future resident was selected as the potential exposure group.

I.7.6. Remediation options and selected approach

The EPA has developed guidance for the investigation and assessment of remedial options, termed a feasibility study, for CERCLA sites [208]. The following steps are included in this process:

- Options are identified and considered;
- Options assessment and/or evaluation is undertaken where applicable;
- Involvement of interested parties is ensured and other stakeholder engagement activities are implemented;
- A decision is made.

[55] Often referred to as the ROD.

The feasibility study included a complete evaluation of the site's hydrogeology, current site conditions, a risk assessment, and alternatives for remedial action. For Maxey Flats, 19 different remediation options were proposed and assessed against each other.

The Record of Decision was issued in September 1991 and defined the remedial solution for the site as follows:

- The installation of an interim cap with periodic replacement, as required;
- Maintenance and monitoring for natural stabilization;
- When stable, construction of a final cap, and monitoring in perpetuity.

Natural stabilization was selected as the preferred remedial approach because, compared with other options, it was the method that was considered least likely to compromise the integrity of disposed containers, such as 55 gallon drums.

In terms of remediation standards (e.g. cleanup criteria, end state criteria, end uses), the maximum tritium contamination level for drinking water to the nearest off-site resident was utilized as the remedial action objective. This option was chosen because the primary contaminant (tritium, with a half-life of 12.3 years) will essentially have decayed in less than 100 years.

The components of the selected remedy in the Record of Decision included the following:

- Demolition of on-site structures.
- Extraction and solidification of approximately 11 million litres of leachate from the trenches.
- Excavation of additional disposal trenches at the site for the disposal of solidified leachate, demolition wastes and site debris.
- Excavation of additional on-site disposal trenches for disposal of site debris and solidified leachate.
- Installation and maintenance of an initial cap consisting of clay and a synthetic liner.
- Relandscaping of the capped disposal area to support the management of surface water runon and runoff.
- Installation of a subsurface barrier to manage groundwater flow, if necessary.
- Installation of a monitoring system to continuously assess infiltration into the trenches and detect the accumulation of leachate, and thus verify the performance of the remediation.
- Implementation of a monitoring regime to assess site water levels and rates of subsidence.

— Sampling and testing of groundwater, surface water and air for a specified range of contaminants.
— Procurement of land adjacent to the site to form a buffer zone and minimize deforestation, and prevent activities that could accelerate hill slope erosion or impact the integrity of the remediation. The buffer zone would also allow for frequent and unrestricted access for monitoring.
— Following the completion of the natural subsidence process, installation of a multilayer engineered cap with a synthetic liner and soil cover.
— Undertaking five year reviews to assess the effectiveness of the remediation and ensure the ongoing achievement of the remedial objectives.
— Implementation of institutional controls to restrict the use of the site and enable maintenance and monitoring in perpetuity [209].

The remedial design for the cap included the use of a geonet for the drainage layer, a 60 mm geomembrane and a geosynthetic clay liner (see Fig. 9).

Interested parties were included in the options evaluation process prior to the final selection of an option. The local stakeholders included homeowners, state environmental groups and workers of the cleanup organizations. Several annual public meetings were held during the decision making process and the implementation of remedial works.

I.7.7. Implementation of the remediation approach

The implementation of remediation was divided into the following four phases:

(a) Initial closure period (22 months);
(b) Interim maintenance period (35 to 100 years);
(c) Final closure period (10 months);
(d) Custodial maintenance period (in perpetuity).

The initial closure period required the following activities and was completed by 2003:

— Installation of an interim geomembrane cover.
— Demolishing site structures and disposing of the waste on-site. Where possible, demolition materials were used for site landscaping activities.
— Extraction of the trench leachate, then solidification with cement into earth-mounded concrete bunkers.
— Construction of an engineered drainage system to manage and minimize rain infiltration into the trenches.

— Subsidence monitoring of the trench area.
— Ongoing groundwater and surface water monitoring.

The Record of Decision assumed that the interim maintenance period of between 35 to 100 years would be required for natural stabilization of the trenches. However, the subsidence monitoring between 2003 and 2013 demonstrated that minimal subsidence had occurred. Groundwater monitoring data were also collected regularly from 2003 in 83 sumps across the site. The statistical analysis of the water level data demonstrated that water levels were not increasing significantly and therefore the installation of a groundwater barrier was not needed in this phase of works.

Therefore, permission was granted by the EPA to move to the final closure period earlier than expected and to proceed with the installation of the permanent earthen cap. The construction of the permanent cap was completed in 2016 and followed the EPA guidance for cover layers [210]. Figure 10 shows the trench site with the completed geomembrane cover.

I.7.8. Conclusions and lessons learned

The project costs have been estimated as US$ 15 million for the site characterization activities and US$ 23 million for the planning and remediation works. It is estimated that a further US$ 10 million will be needed for post-remediation management, which includes long term monitoring and maintenance of the site.

The following lessons learned from the project can be highlighted:

— The benefit of the stakeholder engagement approach used for CERCLA sites enabled the support and agreement of 82 parties in agreeing an end state for the site.
— Understanding the unique groundwater pathways (the bath-tubbing effect), which took many years and iterations to determine, was the key to the final remedy selection. In hindsight, the work may have progressed faster if greater attention had been paid to the groundwater transport pathway earlier on in the assessment.
— Uncertainties still remain regarding the waste inventory. More sampling of the waste prior to the remedial activities may have helped to reduce these uncertainties.
— The site characterization works were impacted by the presence of leachate; therefore, the work to reduce leachate levels ought to have occurred earlier in the work programme.

— The value of planning and implementing through subsidence and water monitoring was realized when the interim maintenance period was significantly reduced.

Appendix II

KEY EXPOSURE PATHWAYS AND SCENARIOS FOR LEGACY TRENCH SITES

II.1. GROUNDWATER PATHWAY

Typically, the most important pathway for legacy trench sites is the groundwater pathway. The pathway is important when the groundwater body at the legacy trench site is used in the following ways:

— Consumed directly as drinking water;
— Used for irrigation of agricultural land, with subsequent consumption of agricultural products;
— Discharged to a surface water body (e.g. lake, river), where a defined reference individual may have direct contact with the contaminated surface water.

Figure 43 shows an example of a conceptual model of the groundwater pathway for post-closure safety assessment of a near surface trench disposal facility to the reference individual [69].

There may be an increase in impact through the groundwater pathway at a legacy trench site due to the lack or degradation of engineered barriers (e.g. bottom liners, caps). There may also be an increased impact via the groundwater pathway where a legacy trench site is situated in hydrogeological settings that do not comply with the modern standards for siting of near surface waste disposal facilities (e.g. a site with a shallow groundwater table and/or having trenches within highly permeable deposits with low sorption capacity). For example, a number of trenches constructed in the Kiev region of Ukraine following the accident at the ChNPP were constructed in sandy soils with a shallow groundwater table [211].

II.2. GROUNDWATER BATH-TUBBING PATHWAY

In some environmental settings, infiltration of rainfall into a legacy trench located in a low permeability geological formation may lead to accumulation of water in the trenches. When this occurs, surface out-flow, or bath-tubbing, can occur. This results in the release of contaminated pore water from the trenches to the soil surface and transport by runoff mechanisms. This type of scenario has

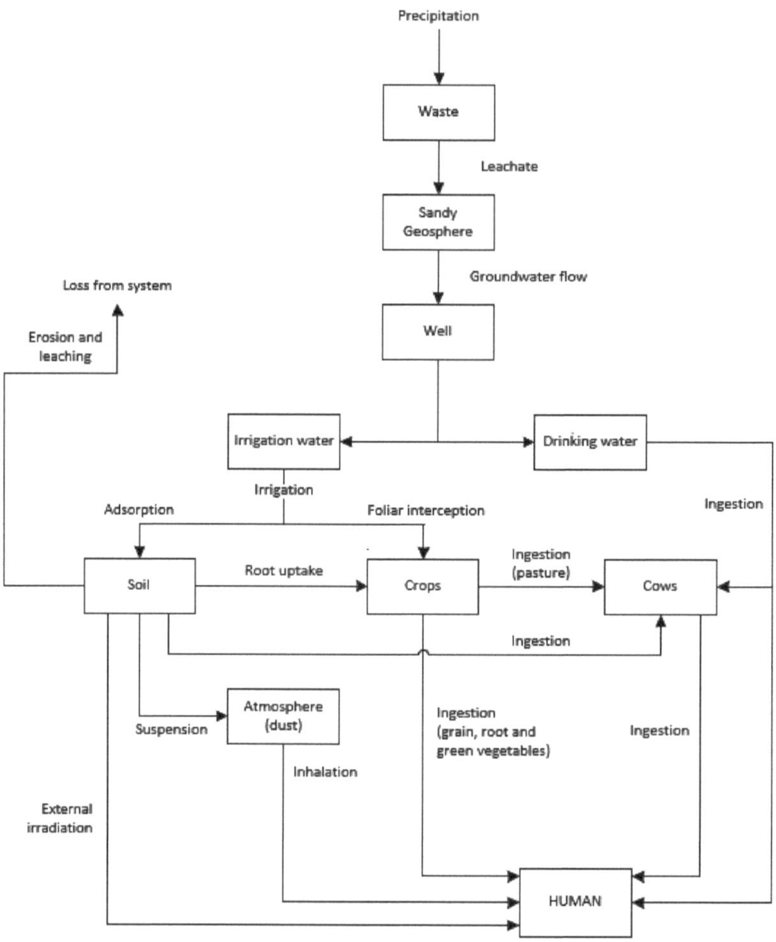

FIG. 43. Conceptual model for the post-closure leaching scenario for a near surface trench disposal facility for the sandy geosphere disposal system. Figure reproduced from Ref. [69].

occurred at the Little Forest site in Australia, leading to soil contamination by plutonium and americium [177], and has also been reported at the Maxey Flats site, USA [212]. The contaminated soil can be the source of secondary exposure of reference individuals (see Fig. 28 and Ref. [213] for further information). For an operational site, this scenario may occur because the trench is open and receiving wastes and there is no cap to minimize infiltration [69].

II.3. DIRECT EXPOSURE

Where the trench cap is absent or degraded and there are inadequate access restrictions, the radioactive wastes at a legacy trench site could pose a risk via direct exposure to the reference individual. The relevant pathways include external exposure, inhalation and occasional ingestion.

Access to a legacy trench site could be accidental, when its location is unknown, or purposeful. Activities such as road construction or forest clearance could result in unintentional intrusion into an unknown legacy trench site. Purposeful access into a legacy trench site may be motivated as an opportunity to exhume scrap metal or other potentially usable materials present in the trench, or as part of a planned activity. An example of this type of direct exposure occurred in 2017 at the Veselovsky Pit legacy trench site in Ukraine. Several radioactively contaminated steel objects and drums were excavated by scrap metal hunters. The site had been forgotten and institutional control lost as a result of administrative changes in the government [214].

II.4. BIOMIGRATION

At unknown legacy trench sites, or sites with insufficient maintenance, vegetation with deep rooting systems may develop over the site. In these instances, vegetation can accumulate contaminants, including radionuclides, from the trench and transport them to the surface. This type of scenario occurred at the Chornobyl Red Forest radioactive waste burial site. In this scenario, the legacy trenches were overgrown by an approximately 30 year old pine forest. The trees accumulated radioactive contaminants (mainly strontium-90 and caesium-137) from legacy trenches. The radionuclides were transported from the trenches to the soil surface via the tree litter fall [196, 215]. Bioaccumulation of radioactivity has resulted in increased external dose rates and risk of atmospheric resuspension of radioactivity in the case of a forest fire.

REFERENCES

[1] BUGAI, D., RODOLFO, A., Scenarios and pathways of radionuclide releases from near-surface waste disposal facilities: A brief overview of historical evidence, Nucl. Radiat. Saf. **87** 3 (2020) 21–27, https://doi.org/10.32918/nrs.2020.3(87).03

[2] INTERNATIONAL ATOMIC ENERGY AGENCY, Remediation Strategy and Process for Areas Affected by Past Activities and Events, IAEA Safety Standard Series No. GSG-15, IAEA, Vienna (2022).

[3] INTERNATIONAL ATOMIC ENERGY AGENCY, Addressing Challenges in Managing Radioactive Waste from Past Activities, IAEA-TECDOC-2039, IAEA, Vienna (2024).

[4] INTERNATIONAL ATOMIC ENERGY AGENCY, Determination of Environmental Remediation End States, IAEA Nuclear Energy Series No. NW-G-3.2, IAEA, Vienna (2021).

[5] UNITED STATES ENVIRONMENTAL PROTECTION AGENCY, Superfund Record of Decision: Maxey Flats Nuclear Disposal, EPA/ROD/R04-91/097, USEPA, Washington, DC (1991).

[6] INTERNATIONAL STANDARDS ORGANISATION, Soil Quality — Sustainable Remediation, ISO 18504:2017, ISO, Geneva (2017).

[7] SUSTAINABLE REMEDIATION FORUM UK, A Framework for Assessing the Sustainability of Soil and Groundwater Remediation — Contaminated Land: Applications in Real Environments, CL:AIRE, London (2010).

[8] EUROPEAN ATOMIC ENERGY COMMUNITY, FOOD AND AGRICULTURE ORGANIZATION OF THE UNITED NATIONS, INTERNATIONAL ATOMIC ENERGY AGENCY, INTERNATIONAL LABOUR ORGANIZATION, OECD NUCLEAR ENERGY AGENCY, PAN AMERICAN HEALTH ORGANIZATION, UNITED NATIONS ENVIRONMENT PROGRAMME, WORLD HEALTH ORGANIZATION, Radiation Protection and Safety of Radiation Sources: International Basic Safety Standards, IAEA Safety Standards Series No. GSR Part 3, IAEA, Vienna (2014).

[9] BUGAI, D., et al., Safety ranking of Chernobyl radioactive legacy sites situated in populated areas for prioritization of remedial measures, Nucl. Phys. At. Energy **20** 1 (2019) 34–43, https://doi.org/10.15407/jnpae2019.01.034

[10] LOW LEVEL WASTE REPOSITORY, The 2011 Environmental Safety Case: Main Report, Report No. LLWR/ESC/R(11)10016, LLW Repository, Holmrook (2011).

[11] INTERNATIONAL COMMISSION ON RADIOLOGICAL PROTECTION, The 2007 Recommendations of the International Commission on Radiological Protection, Publication 103, Elsevier, Amsterdam (2007).

[12] INTERNATIONAL ATOMIC ENERGY AGENCY, Upgrading of Near Surface Repositories for Radioactive Waste, Technical Reports Series No. 433, IAEA, Vienna (2005).

[13] INTERNATIONAL ATOMIC ENERGY AGENCY, Retrieval and Conditioning of Solid Radioactive Waste from Old Facilities, Technical Reports Series No. 456, IAEA, Vienna (2007).

[14] INTERNATIONAL ATOMIC ENERGY AGENCY, Disposal of Radioactive Waste, IAEA Safety Standards Series No. SSR-5, IAEA, Vienna (2011).

[15] INTERNATIONAL ATOMIC ENERGY AGENCY, Near Surface Disposal Facilities for Radioactive Waste, IAEA Safety Standards Series No. SSG-29, IAEA, Vienna (2014).

[16] EUROPEAN ATOMIC ENERGY COMMUNITY, FOOD AND AGRICULTURE ORGANIZATION OF THE UNITED NATIONS, INTERNATIONAL ATOMIC ENERGY AGENCY, INTERNATIONAL LABOUR ORGANIZATION, INTERNATIONAL MARITIME ORGANIZATION, OECD NUCLEAR ENERGY AGENCY, PAN AMERICAN HEALTH ORGANIZATION, UNITED NATIONS ENVIRONMENT PROGRAMME, WORLD HEALTH ORGANIZATION, Fundamental Safety Principles, IAEA Safety Standards Series No. SF-1, IAEA, Vienna (2006).

[17] INTERNATIONAL ATOMIC ENERGY AGENCY, Predisposal Management of Radioactive Waste, IAEA Safety Standards Series No. GSR Part 5, IAEA, Vienna (2009).

[18] INTERNATIONAL ATOMIC ENERGY AGENCY, Decommissioning of Facilities, IAEA Safety Standards Series No. GSR Part 6, IAEA, Vienna (2014).

[19] INTERNATIONAL COMMISSION ON RADIOLOGICAL PROTECTION, Radiological Protection Policy for the Disposal of Radioactive Waste, Publication 77, ICRP, Ottawa (1997).

[20] INTERNATIONAL COMMISSION ON RADIOLOGICAL PROTECTION, Recommendations as Applied to the Disposal of Long-lived Solid Radioactive Waste, Publication 81, ICRP, Ottawa (1998).

[21] PAYNE, T.E., Background Report on the Little Forest Burial Ground Legacy Waste Site, Rep. ANSTO/E-780, ANSTO, Lucas Heights (2012).

[22] TURNER, T.D., BOLLING, D.H., JOHNSON, M.A., Retrieval of Buried Transuranic Waste at Oak Ridge National Laboratory, Proc. 2006 Waste Management Symposium, New York (2006).

[23] INTERNATIONAL ATOMIC ENERGY AGENCY, Knowledge Loss Risk Management in Nuclear Organizations, IAEA Nuclear Energy Series No. NG-T-6.11, IAEA, Vienna (2017).

[24] INTERNATIONAL ATOMIC ENERGY AGENCY, The Nuclear Power Industry's Ageing Workforce: Transfer of Knowledge to the Next Generation, IAEA-TECDOC-1399, IAEA, Vienna (2004).

[25] INTERNATIONAL ATOMIC ENERGY AGENCY, IAEA Safety and Security Glossary, 2022 (Interim Edition), IAEA, Vienna (2022), https://doi.org/10.61092/iaea.rrxi-t56z

[26] CLEVELAND, J.M., REES, T.F., Characterization of Plutonium in Maxey Flats radioactive trench leachates, Science **212** (1981) 1506–1509, https://doi.org/10.1126/science.212.4502.1506

[27] ROWLING, B., et al., Measurement of tributyl phosphate (TBP) in groundwater at a legacy radioactive waste site and its possible role in contaminant mobilisation, J. Environ. Radioact. **178–179** (2017) 377–384, https://doi.org/10.1016/j.jenvrad.2017.05.015

[28] FELLINGHAM, L., GRAHAM, A., STIFF, S. (Eds), "Characterisation and remediation of beryllium waste pits in the southern storage area at Harwell", ASME 2003, 9th Int. Conf. Radioactive Waste Management and Environmental Remediation (2003), https://doi.org/10.1115/ICEM2003-4861

[29] INTERNATIONAL STANDARDS ORGANISATION, Soil Quality — Conceptual Site Models for Potentially Contaminated Sites, ISO 21365:2019, ISO, Geneva (2019).

[30] CANADIAN COUNCIL OF MINISTERS OF THE ENVIRONMENT, Subsurface Assessment Handbook for Contaminated Sites, Rep. EPC-NCSRP-48E, CCME, Winnipeg (1994).

[31] INTERNATIONAL STANDARDS ORGANISATION, Soil quality — Sampling — Part 202: Preliminary Investigations, ISO 18400-2002:2018, ISO, Geneva (2018).

[32] INTERNATIONAL ATOMIC ENERGY AGENCY, Retrieval, Restoration and Maintenance of Old Radioactive Waste Inventory Records, IAEA-TECDOC-1548 IAEA, Vienna (2007).

[33] INTERNATIONAL ATOMIC ENERGY AGENCY, Strategy and Methodology for Radioactive Waste Characterization, IAEA-TECDOC-1537, IAEA, Vienna (2007).

[34] INTERNATIONAL ATOMIC ENERGY AGENCY, Management of Long Term Radiological Liabilities: Stewardship Challenges, Technical Report Series No. 450, IAEA, Vienna (2006).

[35] INTERNATIONAL ATOMIC ENERGY AGENCY, Characterization of Radioactive Waste Forms and Packages, Technical Reports Series No. 383, IAEA, Vienna (1997).

[36] INTERNATIONAL ATOMIC ENERGY AGENCY, Categorizing Operational Radioactive Wastes, IAEA-TECDOC-1538, IAEA, Vienna (2007).

[37] INTERNATIONAL STANDARDS ORGANISATION, Soil quality — Sampling — Part 104: Strategies, ISO 18400-104:2018, ISO, Geneva (2018).

[38] INTERNATIONAL STANDARDS ORGANISATION, Soil quality — Characterization of Contaminated Soil Related to Groundwater Protection, ISO 15175:2018, ISO, Geneva (2018).

[39] CANADIAN COUNCIL OF MINISTERS OF THE ENVIRONMENT, Guidance Manual for Environmental Site Characterization in support of Environmental and Human Health Risk Assessment, Vol. 1, Rep. PN 1551, CCME, Winnipeg (2016).

[40] INTERNATIONAL ATOMIC ENERGY AGENCY, Characterization of Radioactively Contaminated Sites for Remediation Purposes, IAEA-TECDOC-1017, IAEA, Vienna (1998).

[41] TOWLER, P., RANKINE, A., KRUSE, P., ESLAVA-GOMEZ, A., SAFEGROUNDS — Good Practice Guidance for Site Characterisation, Version 2, CIRIA, London (2009).

[42] UNITED STATES NUCLEAR REGULATORY COMMISSION, UNITED STATES ENVIRONMENTAL PROTECTION AGENCY, UNITED STATES DEPARTMENT OF ENERGY, UNITED STATES DEPARTMENT OF DEFENSE, Multi-Agency Radiation Survey and Site Investigation Manual (MARSSIM) Rev. 1, Rep. NUREG-1575, Rev. 1; EPA-402-R-97-016, Rev. 1; DOE/EH-0624, Rev. 1, NRC, Washington, DC (2000).

[43] INTERNATIONAL STANDARDS ORGANISATION, Measurement of Radioactivity in the Environment – Soil – Part 2: Guidance for the Selection of the Sampling Strategy, Sampling and Pre-treatment of Samples, ISO 18589-2:2015, ISO, Geneva (2015).

[44] INTERNATIONAL STANDARDS ORGANISATION, Soil quality — Sampling — Part 101: Framework for the Preparation and Application of a Sampling Plan, ISO 18400-101:2017, ISO, Geneva (2017).

[45] UNITED STATES ENVIRONMENTAL PROTECTION AGENCY, Guidance on Systematic Planning Using the Data Quality Objectives Process, EPA QA/G-4, US EPA, Washington, DC (2006).

[46] INTERNATIONAL ATOMIC ENERGY AGENCY, In Situ Analytical Characterization of Contaminated Sites Using Nuclear Spectrometry Techniques, Analytical Quality in Nuclear Applications Series No. 49, IAEA, Vienna (2017).

[47] KEAREY, P., BROOKS, M., HILL, I., An Introduction to Geophysical Exploration, 3rd edn, Blackwell Science, Oxford (2002).

[48] McDOWELL, P.W., et al., Geophysics in Engineering Investigations, Geological Society of London, London (2002).

[49] INTERNATIONAL STANDARDS ORGANISATION, Hydrometry — Groundwater — Surface Geophysical Surveys for Hydrogeological Purposes, ISO/TR 21414:2016, ISO, Geneva (2016).

[50] OECD NUCLEAR ENERGY AGENCY, Nuclear Site Remediation and Restoration During Decommissioning of Nuclear Installations, Rep. 7192, NEA, Paris (2014).

[51] PAYNE, T.E., et al., Installation of a Pilot Experimental Trench at the Little Forest Legacy Site, Rep. ANSTO-E-786, ANSTO, Lucas Heights (2018).

[52] HANKIN, S., Little Forest Burial Ground – Geology, Geophysics and Well Installation 2009–2010, Vols 1 & 2, Rep. ANSTO-E-781, ANSTO, Lucas Heights (2013).

[53] INTERNATIONAL STANDARDS ORGANISATION, Soil quality — Sampling — Part 102: Selection and Application of Sampling Techniques, ISO 18400-102:2017, ISO, Geneva (2017).

[54] INTERNATIONAL ATOMIC ENERGY AGENCY, Guidelines on Soil and Vegetation Sampling for Radiological Monitoring, Technical Report Series No. 486, IAEA, Vienna (2019).

[55] STENHOUSE, M.J., KIRKO, V.I. (Eds), Defence Nuclear Waste Disposal in Russia: International Perspective, Springer Science & Business Media, Dordrecht (1998), https://doi.org/10.1007/978-94-011-5112-2

[56] LOW LEVEL WASTE REPOSITORY, The 2011 Environmental Safety Case: Hydrogeology, Rep. LLWR/ESC/R(11)10022, LLW Repository, Holmrook (2011).

[57] INTERNATIONAL ATOMIC ENERGY AGENCY, The Safety Case and Safety Assessment for the Disposal of Radioactive Waste, IAEA Safety Standards Series No. SSG-23, IAEA, Vienna (2012).

[58] GOLDAMMER, W., NÜSSER, A., BÜTOW, E., LÜHR, H.-P., "Integrated assessment of radiological and non-radiological risks at contaminated sites", Mathematical Methods Applied to Environmental Investigations (THIERGÄRTNER, H., Ed.), CPress Publishing House, Dresden (1999) 53–72.

[59] SIMON, T., Environmental Risk Assessment: A Toxicological Approach, 2nd edn, CRC Press, Boca Raton, FL (2020),
https://doi.org/10.1201/9780429286001

[60] RAZZAGHI, M., Statistical Models in Toxicology, 1st edn, Chapman and Hall/CRC, New York (2020),
https://doi.org/10.1201/9780429155185-1

[61] BUGAI, D., et al., Geochemical influence of waste trench no. 22T at Chernobyl Pilot Site at the aquifer: Long-term trends, governing processes, and implications for radionuclide migration, Appl. Geochem. **27** 7 (2012) 1320–1338,
https://doi.org/10.1016/j.apgeochem.2011.09.021

[62] TOROPOV, A.S., Migration forms of anthropogenic radionuclides in tunnel waters at the Degelen Mountains, Semipalatinsk test site, Geochem. Int. **58** 3 (2020) 342–351,
https://doi.org/10.1134/S0016702920020123

[63] KERSTING, A., "Impact of colloidal transport on radionuclide migration in the natural environment", Radionuclide Behaviour Natural Environment: Science, Implications and Lessons for the Nuclear Industry, Woodhead Publishing, Cambridge (2012) 384–410,
https://doi.org/10.1533/9780857097194.2.384

[64] SAUNDERS, J.A., TORAN, L.E., Modeling of radionuclide and heavy metal sorption around low- and high-pH waste disposal sites at Oak Ridge, Tennessee, Appl. Geochem. **10** 6 (1995) 673–684,
https://doi.org/10.1016/0883-2927(95)00036-4

[65] INTERNATIONAL ATOMIC ENERGY AGENCY, Safety Assessment Methodologies for Near Surface Disposal Facilities, Vol. 1, IAEA, Vienna (2004).

[66] BATANDJIEVA, B., ASAM – The International Programme on Application of Safety Assessment Methodologies for Near Surface Radioactive Waste Disposal Facilities, International Conference on Issues and Trends in Radioactive Waste Management, Vienna, Rep. IAEA-CN-90, IAEA, Vienna (2002).

[67] INTERNATIONAL ATOMIC ENERGY AGENCY, Safety Assessment Methodologies, Coordinated Research Project on the Application of Safety Assessment Methodologies for Near-Surface Radioactive Waste Disposal Facilities (ASAM), Regulatory Review Working Group Report, IAEA, Vienna (2002).

[68] GUSKOV, A., BATANDJIEVA, B., KOZAK, M.W., TORRES-VIDAL, C., RADON-type Disposal Facility Safety Case for the Co-coordinated Research Project on Improvement of Safety Assessment Methodologies for Near Surface Radioactive Waste Disposal Facilities (ISAM), Int. Conf. Issues and Trends in Radioactive Waste Management, Vienna, Rep. IAEA-CN-90/72, IAEA, Vienna (2002).

[69] INTERNATIONAL ATOMIC ENERGY AGENCY, Derivation of Activity Limits for the Disposal of Radioactive Waste in Near Surface Disposal Facilities, IAEA-TECDOC-1380, IAEA, Vienna (2004).

[70] BIOPROTA, BIOMASS 2020: Interim Report, Svensk Kärnbränslehantering AB, Solna (2018).

[71] KONIKOW, L.F., MERCER, J.W., Groundwater flow and transport modeling. J. Hydrol. **100** 1–3 (1988) 379–409, https://doi.org/10.1016/0022-1694(88)90193-X

[72] INTERNATIONAL ATOMIC ENERGY AGENCY, Technical Options for the Remediation of Contaminated Groundwater, IAEA-TECDOC-1088, IAEA, Vienna (1999).

[73] MIDDLEMIS, H., Groundwater Flow Modelling Guideline, Murray-Darling Basin Commission, Perth (2001).

[74] BARNETT, B., et al., Australian Groundwater Modelling Guidelines, Rep. No. 82, Australian Government National Water Commission, Canberra (2012).

[75] INTERNATIONAL ATOMIC ENERGY AGENCY, Environmental Modelling for Radiation Safety (EMRAS): A Summary Report of the Results of the EMRAS Programme (2003–2007), IAEA-TECDOC-1678, IAEA, Vienna (2012).

[76] HAVERKAMP, B., KROEGER, H. (Eds), Safety assessment for the radon-type surface disposal facility at Saakadze, Georgia – 15406, WM2015: Annual Waste Management Symposium, Phoenix, AZ (2015), Tempe (2015).

[77] LOW LEVEL WASTE REPOSITORY, The 2011 Environmental Safety Case: Assessment of Long-term Radiological Impacts, Rep. LLWR/ESC/R(11)10028, LLW Repository, Holmrook (2011).

[78] JOHANSEN, M.P., et al., Little Forest Legacy Site — Technical Report: Dose Rate Estimates to Humans and Wildlife for a Range of Potential Future Scenarios, Rep. ANSTO-E-787, ANSTO, Lucas Heights (2020).

[79] AMERICAN SOCIETY FOR TESTING AND MATERIALS, Standard Guide for Subsurface Flow and Transport Modeling, Rep. ASTM D5880-95(2000), ASTM International, West Conshohocken, PA (2017).

[80] AMERICAN SOCIETY FOR TESTING AND MATERIALS, Standard Guide for Conducting a Sensitivity Analysis for a Groundwater Flow Model Application, Rep. ASTM D5611-94(2016), ASTM International, West Conshohocken, PA (2016).

[81] AMERICAN SOCIETY FOR TESTING AND MATERIALS, Standard Guide for Defining Initial Conditions in Groundwater Flow Modeling, Rep. ASTM D5610-94(2014), ASTM International, West Conshohocken, PA (2014).

[82] AMERICAN SOCIETY FOR TESTING AND MATERIALS, Standard Guide for Comparing Groundwater Flow Model Simulations to Site-Specific Information, Rep. ASTM D5490-93(2014)e1, ASTM International, West Conshohocken, PA (2014).

[83] AMERICAN SOCIETY FOR TESTING AND MATERIALS, Standard Guide for Documenting a Groundwater Flow Model Application, Rep. ASTM D5718-13, ASTM International, West Conshohocken, PA (2013).

[84] AMERICAN SOCIETY FOR TESTING AND MATERIALS, Standard Guide for Calibrating a Groundwater Flow Model Application, Rep. ASTM D5981/D5981M-18, ASTM International, West Conshohocken, PA (2018).

[85] DUNNIVANT, F.M., ANDERS, E., Pollutant Fate and Transport in Environmental Multimedia, Wiley, Hoboken, NJ (2019).

[86] PETERS, G., SVANSTRÖM, M., Environmental Sustainability for Engineers and Applied Scientists, Cambridge University Press, Cambridge (2019), https://doi.org/10.1017/9781316711408

[87] BEAR, J., CHENG, A.H.-D., Modeling Groundwater Flow and Contaminant Transport, Springer Netherlands, Dordrecht (2010), https://doi.org/10.1007/978-1-4020-6682-5

[88] CHU, M.S.Y., KOZAK, M.W., CAMPBELL, J.E., THOMPSON, B.H., A Self-teaching Curriculum for the NRC/SNL (Nuclear Regulatory Commission/Sandia National Laboratory) Low-level Waste Performance Assessment Methodology, Rep. SAND–90-0585, Nuclear Regulatory Commission, Washington, DC (1991), https://doi.org/10.2172/6049453

[89] HUGHES, J.D., LANGEVIN, C.D., BANTA, E.R., Documentation for the MODFLOW 6 Framework, Rep. 6-A57, Reston, VA (2017), https://doi.org/10.3133/tm6A57

[90] ZHENG, C., WANG, P.P., MT3DMS: A Modular Three-Dimensional Multispecies Transport Model for Simulation of Advection, Dispersion, and Chemical Reactions of Contaminants in Groundwater Systems; Documentation and User's Guide, Rep. SERDP-99-1, US Army Corps of Engineers, Washington, DC (1999).

[91] CONNECTFLOW TEAM, ConnectFlow Technical Summary, Rep. Version 12.2, Jacobs, Didcot (2020).

[92] BUGAI, D.A., WATERS, R.D., DZHEPO, S.P., SKAL'SKIJ, A.S., Risks from radionuclide migration to groundwater in the Chernobyl 30-km zone, Health Phys. **71** 1 (1996) 9–18, https://doi.org/10.1097/00004032-199607000-00002

[93] BUGAI, D., et al., Radionuclide migration at experimental polygon at Red Forest waste site in Chernobyl Zone; Part 2: Hydrogeological characterization and groundwater transport modeling, Appl. Geochem. **27** 7 (2012) 1359–1374, https://doi.org/10.1016/j.apgeochem.2011.09.028

[94] MOLITOR, N., et al., Recent safety Assessment Findings on Management of Legacy Wastes from Chornobyl Accident, Int. Conf. on Nuclear Decommissioning and Environment Recovery INUDECO 2017, Slavutych, Ukraine (2017).

[95] NUCLEAR DECOMMISSIONING AUTHORITY, Strategy Effective from April 2021, NDA, Moor Row (2021).

[96] NUCLEAR DECOMMISSIONING AUTHORITY, Strategy Effective from April 2016, NDA, Moor Row (2016).

[97] NUCLEAR DECOMMISSIONING AUTHORITY, The NDA Value Framework, Version 1.2, NDA, Moor Row (2016).

[98] UNITED STATES DEPARTMENT OF ENERGY, Guidebook to Decision-Making Methods, WSRC-IM-2002-00002, US DOE, Washington, DC (2002).

[99] DEPARTMENT OF BUSINESS ENERGY & INDUSTRIAL STRATEGY, Amending the Framework for the Final Stages of Nuclear Decommissioning and Clean-up Government Response to Consultation, BEIS, London (2018).

[100] DEPARTMENT OF BUSINESS ENERGY & INDUSTRIAL STRATEGY, Nuclear Decommissioning Consultation of the Regulation of Nuclear Sites in the Final Stages of Decommissioning and Clean-Up, BEIS, London (2018).

[101] INTERNATIONAL ATOMIC ENERGY AGENCY, Non-technical Factors Impacting on the Decision Making Processes in Environmental Remediation, IAEA-TECDOC-1279, IAEA, Vienna (2002).

[102] NUCLEAR INDUSTRY SAFETY DIRECTORS FORUM, Best Available Techniques (BAT) for the Management of the Generation and Disposal of Radioactive Wastes: A Nuclear Industry Code of Practice (2010), https://nuclearinst.com/EARWG-

[103] PENFOLD, J., SAFEGROUNDS: Guide to the Comparison of Contaminated Land Management Options, W28, CIRIA, London (2009).

[104] DEPARTMENT FOR COMMUNITIES AND LOCAL GOVERNMENT, Multi-criteria Analysis: A Manual, Communities and Local Government Publications, London (2009).

[105] BARDOS, R.P., LEWIS, A.J., NORTCLIFF, S., MARIOTTI, C., SULLIVAN, T., Review of Decision Support Tools for Contaminated Land Management, and Their Use in Europe, Final Report, Austrian Federal Environment Agency, Vienna (2002).

[106] MARCOMINI, A., SUTER, G.W. II, CRITTO, A. (Eds), Decision Support Systems for Risk-Based Management of Contaminated Sites, Springer US, New York (2009), https://doi.org/10.1007/978-0-387-09722-0

[107] SMITH, J.W.N., KERRISON, G., Benchmarking of decision-support tools used for tiered sustainable remediation appraisal, Water, Air, Soil Pollut. **224** 12 (2013) 1706, https://doi.org/10.1007/s11270-013-1706-y

[108] HUYSEGOMS, L., CAPPUYNS, V., Critical review of decision support tools for sustainability assessment of site remediation options, J. Environ. Manage. **196** (2017) 278–296, https://doi.org/10.1016/j.jenvman.2017.03.002

[109] SULLIVAN, T., BARDOS, P., MAROT, C., MARIOTTI, R., Framework for decision support used in contaminated land management in Europe and North America, Land Contam. Reclam. **9** (2001).

[110] NATIONAL RESEARCH COUNCIL, Environmental Cleanup at Navy Facilities: Adaptive Site Management, The National Academies Press, Washington, DC (2003) 376.

[111] PRICE, J., SPRENG, C., HAWLEY, E.L., DEEB, R., Remediation management of complex sites using an adaptive site management approach, J. Environ. Manage. **204** (2017) 738–747, https://doi.org/10.1016/j.jenvman.2017.04.009

[112] INTERSTATE TECHNOLOGY & REGULATORY COUNCIL, Remediation Management of Complex Sites Washington: ITRC (2017), https://rmcs-1.itrcweb.org

[113] INTERNATIONAL ATOMIC ENERGY AGENCY, Remediation of Sites with Dispersed Radioactive Contamination, Technical Report Series No. 424, IAEA, Vienna (2004).

[114] UNITED STATES ENVIRONMENTAL PROTECTION AGENCY, Technology Reference Guide for Radioactively Contaminated Media, Rep. EPA 402-R-07-004, US EPA, Washington, DC (2007).

[115] MALLETT, H., SAFEGROUNDS: Technical Options for Managing Contaminated Land, W18, CIRIA, London (2004).

[116] RUDLAND, D.J., JACKSON, S.D., Selection of Remedial Treatments for Contaminated Land – A Guide to Good Practice, C622, CIRIA, London (2004).

[117] EVANS, D., JEFFERIS, S.A., THOMAS, A.O., CUI, S., Remedial Processes for Contaminated Land: Principles and Practice, C549, CIRIA, London (2001).

[118] SIMON, F.-G., MEGGYES, T., "Sustainable remediation methods for metals and radionuclides", Encyclopedia of Sustainability Science and Technology (MEYERS, R.A., Ed.), Springer, New York (2012) 10403–10435, https://doi.org/10.1007/978-1-4419-0851-3_63

[119] UNITED STATES DEPARTMENT OF ENERGY, 2010 Remediation Effectiveness Report for the U.S. Department of Energy Oak Ridge Reservation, Oak Ridge, Tennessee, Rep. DOE/OR/01-2437&D2, US DOE, (2010).

[120] STATE AGENCY OF UKRAINE ON EXCLUSION ZONE MANAGEMENT. Removal of Radioactive Waste from the Pisky-1 Facility Started in Ivankiv District State Agency of Ukraine on Exclusion Zone Management (2019), https://www.kmu.gov.ua/news/v-ivankivskomu-rajoni-rozpochato-viluchennya-radioaktivnih-vidhodiv-z-obyekta-piski-1

[121] BUGAI, D., GEBAUER, J., SIZOV, A., MOLITOR, N., Practical approach for assessment of end-state radiological criteria for remediation of radioactively contaminated sites, Nucl. Radiat. Saf. **81** 1 (2019) 36–41, https://doi.org/10.32918/nrs.2019.1(81).06

[122] INTERNATIONAL ATOMIC ENERGY AGENCY, Procedures and Techniques for Closure of Near Surface Disposal Facilities for Radioactive Waste, IAEA-TECDOC-1260, IAEA, Vienna (2001).

[123] UNITED STATES ENVIRONMENTAL PROTECTION AGENCY, Third Five-Year Review Report for Maxey Flats Disposal Site, US EPA, Washington, DC (2012).

[124] BLANKENHORN, J., BOWER, B., West Valley Demonstration Project — Past, Present and Future, Waste Management Symposium, 2011, Phoenix, AZ (2011).

[125] LOW LEVEL WASTE REPOSITORY, The 2011 Environmental Safety Case: Engineering Design, Rep. LLWR/ESC/R(11)10020, LLW Repository, Holmrook (2011).

[126] JOHNSON, C., CANGE, J., SKINNER, R., ADAMS, V. (Eds), Remediation of the Melton Valley Watershed at Oak Ridge National Lab: An Accelerated Closure Success Story, Waste Management Symposium, 2008, Phoenix, AZ (2008).

[127] UNITED STATES DEPARTMENT OF ENERGY, 2019 Remediation Effectiveness Report for the U.S. Department of Energy, Oak Ridge Site, Oak Ridge, Tennessee, Rep. DOE/OR/01-2787&D1, Oak Ridge LLC, TN (2019).

[128] FRANCIS, C.W., SPENCE, R.D., TAMURA, T., SPALDING, B.P., In Situ Grouting of Low-Level Burial Trenches with a Cement-Based Grout at Oak Ridge National Laboratory, United States, Rep. ORNL/TM-11838, ORNL, Oak Ridge, TN (1993),
https://doi.org/10.2172/10145965

[129] UNITED STATES ENVIRONMENTAL PROTECTION AGENCY, EPA Superfund Record of Decision: Declaration for Organic Contamination in the Vadose Zone Idaho National Engineering Laboratory, Rep. EPA/ROD/R10-95/086, US EPA, Washington, DC (1995).

[130] ARENAZ, M.R. (Ed.), Remediation of Buried Waste at the Idaho National Laboratory Site, Waste Management Symposium, 2010, Phoenix, AZ (2010).

[131] HUTCHISON, D.P., WEBBER, F.L. (Eds), Balancing Cleanup and Future Land Use at the Idaho National Laboratory, Waste Management Symposium, 2011, Phoenix, AZ (2011).

[132] MARALINGA REHABILITATION TECHNICAL ADVISORY COMMITTEE, Rehabilitation of Former Nuclear Test Sites at Emu and Maralinga (Australia), Department of Education Science and Training, Canberra (2003).

[133] CANADIAN NUCLEAR LABORATORIES, Environmental Risk Assessment of Chalk River Laboratories, Rep. ENVP-509220-REPT-003, Canadian Nuclear Laboratories, Chalk River (2019).

[134] INTERNATIONAL ATOMIC ENERGY AGENCY, Technologies for Remediation of Radioactively Contaminated Sites, IAEA-TECDOC-1086, IAEA, Vienna (1999).

[135] TRUEX, M., et al., Performance assessment of pump-and-treat systems. Groundwater Monit. R. **37** (2017) 28–44,
https://doi.org/10.1111/gwmr.12218

[136] UNITED STATES ENVIRONMENTAL PROTECTION AGENCY, Remediation Technologies,
https://clu-in.org/remediation/

[137] UNITED STATES DEPARTMENT OF DEFENCE, UNITED STATES ENVIRONMENTAL PROTECTION AGENCY, UNITED STATES DEPARTMENT OF ENERGY, ENVIRO.wiki,
http://www.enviro.wiki/index.php?title=Main_Page

[138] REDDY, K.R., CAMESELLE, C. (Eds), Electrochemical Remediation Technologies for Polluted Soils, Sediments and Groundwater, John Wiley & Sons, Hoboken, NJ (2009),
https://doi.org/10.1002/9780470523650

[139] GUPTA, D.K., WALTHER, C. (Eds), Radionuclide Contamination and Remediation Through Plants, Springer, Cham (2014),
https://doi.org/10.1007/978-3-319-07665-2

[140] SINGH, S.N., TRIPATHI, R.D. (Eds), Environmental Bioremediation Technologies, Springer, New York (2007),
https://doi.org/10.1007/978-3-540-34793-4

[141] GUPTA, D.K., SCHULZ, W., STEINHAUSER, G., WALTHER, C., Radiostrontium transport in plants and phytoremediation. Environ. Sci. Pollut. Res. **25** 30 (2018) 29996–30008,
https://doi.org/10.1007/s11356-018-3088-6
[142] CAREY, M.A., FINNAMORE, J.R., MORREY, M.J., MARSLAND, P.A., Guidance on the Assessment and Monitoring of Natural Attenuation of Contaminants in Groundwater, R&D Publication 95, Environment Agency, Bristol (2000).
[143] CONTAMINATED LAND: APPLICATIONS IN REAL ENVIRONMENTS, Understanding Soil Washing, Technical Bulletin TB13, CL:AIRE, London (2017).
[144] BARNETT, F., LYNN, S., REISMANN, D., Technology Performance Review: Selecting and Using Solidification/Stabilization Treatment for Site Remediation, US EPA, Washington, DC (2009).
[145] SIMON, F.-G., MEGGYES, T., "Sustainable remediation methods for radionuclides", Encyclopedia of Sustainability Science and Technology (MEYERS, R.A., Ed.), Springer, New York (2012) 10403–10435,
https://doi.org/10.1007/978-1-4419-0851-3_63
[146] INTERSTATE TECHNOLOGY & REGULATORY COUNCIL, Permeable Reactive Barrier: Technology Update, PRB-5, ITRC, Washington, DC (2011).
[147] INTERNATIONAL ATOMIC ENERGY AGENCY, Communication with the Public in a Nuclear or Radiological Emergency, IAEA, Vienna (2012).
[148] INTERNATIONAL ATOMIC ENERGY AGENCY, Advancing Implementation of Decommissioning and Environmental Remediation Programmes, IAEA Nuclear Energy Series No. NW-T-1.10, IAEA, Vienna (2016).
[149] INTERNATIONAL ATOMIC ENERGY AGENCY, Stakeholder Involvement in Nuclear Issues, INSAG-20, IAEA, Vienna (2006).
[150] INTERNATIONAL ATOMIC ENERGY AGENCY, Communication and Stakeholder Involvement in Environmental Remediation Projects, IAEA Nuclear Energy Series No. NW-T-3.5, IAEA, Vienna (2014).
[151] INTERNATIONAL ATOMIC ENERGY AGENCY, Communication and Consultation with Interested Parties by the Regulatory Body, IAEA Safety Standards Series No. GSG-6, IAEA, Vienna (2017).
[152] INTERNATIONAL ATOMIC ENERGY AGENCY, Stakeholder Involvement Throughout the Life Cycle of Nuclear Facilities, IAEA Nuclear Energy Series No. NG-T-1.4, IAEA, Vienna (2011).
[153] OECD NUCLEAR ENERGY AGENCY, Fostering a Durable Relationship Between a Waste Management Facility and its Host Community, Radioactive Waste Management NEA No. 7264, OECD Publishing, Paris (2015).
[154] OECD NUCLEAR ENERGY AGENCY, Stakeholder Confidence in Radioactive Waste Management: An Annotated Glossary of Key Terms, Radioactive Waste Management NEA No. 6988, OECD Publishing, Paris (2013).
[155] OECD NUCLEAR ENERGY AGENCY, Stakeholder Involvement in Decision Making: A Short Guide to Issues, Approaches and Resources, Radioactive Waste Management NEA No. 7189, OECD Publishing, Paris (2016).

[156] COVELLO, V., SANDMAN, P.M., Risk Communication: Evolution and Revolution (WOLBARST, A.B. Ed.), John Hopkins University Press, Baltimore (2001).

[157] UNITED STATES DEPARTMENT OF ENERGY OFFICE OF LEGACY MANAGEMENT, Fernald Preserve, Ohio Visitors Center, https://www.energy.gov/lm/fernald-preserve-visitors-center

[158] INTERNATIONAL ATOMIC ENERGY AGENCY, Methods for Maintaining a Record of Waste Packages During Waste Processing and Storage, Technical Reports Series No. 434, IAEA, Vienna (2005).

[159] OECD NUCLEAR ENERGY AGENCY, Preservation of Records, Knowledge and Memory (RK&M) Across Generations: Developing a Key Informaiton File for a Radioactive Waste Repository, Radioactive Waste Management NEA No. 7421, OECD Publishing, Paris (2019).

[160] INTERNATIONAL ATOMIC ENERGY AGENCY, Long Term Preservation of Information for Decommissioning Projects, Technical Reports Series No. 467, IAEA, Vienna (2008).

[161] OECD NUCLEAR ENERGY AGENCY, Preservation of Records, Knowledge and Memory Across Generations: Final Report, Radioactive Waste Management NEA No. 7421, OECD Publishing, Paris (2019).

[162] INTERNATIONAL ATOMIC ENERGY AGENCY, Maintenance of Records for Radioactive Waste Disposal, IAEA-TECDOC-1097, IAEA, Vienna (1999).

[163] INTERNATIONAL ATOMIC ENERGY AGENCY, Records for Radioactive Waste Management up to Repository Closure: Managing the Primary Level Information (PLI) Set, IAEA-TECDOC-1398, IAEA, Vienna (2004).

[164] INTERNATIONAL ATOMIC ENERGY AGENCY, Record Keeping for the Decommissioning of Nuclear Facilities, Technical Reports Series No. 411, IAEA, Vienna (2003).

[165] INTERNATIONAL ATOMIC ENERGY AGENCY, Knowledge Management and Its Implementation in Nuclear Organizations, IAEA Nuclear Energy Series No. NG-T-6.10, IAEA, Vienna (2016).

[166] INTERNATIONAL ATOMIC ENERGY AGENCY, Comparative Analysis of Methods and Tools for Nuclear Knowledge Preservation, IAEA Nuclear Energy Series No. NG-T-6.7, IAEA, Vienna (2011).

[167] NUCLEAR DECOMMISSIONING AUTHORITY, Information Governance Strategy, SMS/TS/F3-IKM/001/C, NDA, Moor Row (2013).

[168] NUCLEAR DECOMMISSIONING AUTHORITY, Case Study NDA Archive: Nucleus (the Nuclear and Caithness Archives) (2015), https://www.gov.uk/government/case-studies/nda-archive

[169] PAYNE, T.E., Little Forest Legacy Site: Summary of Site History Until the Commencement of Waste Disposal in 1960, Rep. ANSTO;/E-782, ANSTO, Lucas Heights (2015).

[170] PAYNE, T.E., SHATWELL, J., COMARMOND, M.J., Waste Disposals at the Little Forest Site: Procedures, Records and Significant Events, Rep. ANSTO/E-788, ANSTO, Lucas Heights (2022).

[171] JOINT CONVENTION ON THE SAFETY OF SPENT FUEL MANAGEMENT AND ON THE SAFETY OF RADIOACTIVE WASTE MANAGEMENT, National Report Argentine Republic, Comisión Nacional de Energía Atómica, Buenos Aires (2005).

[172] JOINT CONVENTION ON THE SAFETY OF SPENT FUEL MANAGEMENT AND ON THE SAFETY OF RADIOACTIVE WASTE MANAGEMENT, National Report, Third National Report Argentine Republic, Comisión Nacional de Energía Atómica, Buenos Aires (2008).

[173] JOINT CONVENTION ON THE SAFETY OF SPENT FUEL MANAGEMENT AND ON THE SAFETY OF RADIOACTIVE WASTE MANAGEMENT, National Report, Fifth National Report Argentine Republic, Comisión Nacional de Energía Atómica, Buenos Aires (2014).

[174] JOINT CONVENTION ON THE SAFETY OF SPENT FUEL MANAGEMENT AND ON THE SAFETY OF RADIOACTIVE WASTE MANAGEMENT, National Report Argentine Republic, Comisión Nacional de Energía Atómica, Buenos Aires (2003).

[175] AUSTRALIAN ATOMIC ENERGY COMMISSION, Little Forest Burial Ground: An Information Paper, Rep. AAEC-DR-19, ANSTO, Lucas Heights (1985).

[176] ISAACS, S.R., MEARS, K.F., Study of the Burial Ground Used for Radioactive Waste at the Little Forest Area Near Lucas Heights, New South Wales, Rep. E-247, Australian Atomic Energy Commission, Lucas Heights (1977).

[177] PAYNE, T.E., et al., Trench 'bathtubbing' and surface plutonium contamination at a legacy radioactive waste site, Environ. Sci. Technol. **47** 23 (2013) 13284–13293, https://doi.org/10.1021/es403278r

[178] HUGHES, C.E., et al., Movement of a tritium plume in shallow groundwater at a legacy low-level radioactive waste disposal site in eastern Australia, J. Environ. Radioact. **102** 10 (2011) 943–952,
https://doi.org/10.1016/j.jenvrad.2010.05.009

[179] CENDÓN, D.I., et al., Identification of sources and processes in a low-level radioactive waste site adjacent to landfills: groundwater hydrogeochemistry and isotopes, Aust. J. Earth Sci. **62** 1 (2015) 123–141,
https://doi.org/10.1080/08120099.2015.975155

[180] IKEDA-OHNO, A., et al., Solution speciation of plutonium and americium at an Australian legacy radioactive waste disposal site, Environ. Sci. Technol. **48** 17 (2014) 10045–10053,
https://doi.org/10.1021/es500539t

[181] VÁZQUEZ-CAMPOS, X., et al., Response of microbial community function to fluctuating geochemical conditions within a legacy radioactive waste trench environment, Appl. Environ. Microbiol. **83** 17 (2017) e00729-17,
https://doi.org/10.1128/AEM.00729-17

[182] KINSELA, A.S., et al., Influence of dissolved silicate on rates of Fe(II) oxidation, Environ. Sci. Technol. **50** 21 (2016) 11663–11671,
https://doi.org/10.1021/acs.est.6b03015

[183] CHALK RIVER LABORATORIES, A Description of the Environmental Baseline for Environmental Assessments, Rep. CRL-509200-ENA-001, Atomic Energy of Canada Ltd, Chalk River (2008).

[184] ZULOAGA, P., et al., Co-operation in the development of a policy and strategy for the management of spent nuclear fuel (including provisions for its safe interim storage) and radioactive waste in Mexico, Nucl. Esp. **2013** (1996) 47–51.
[185] DOMÍNGUEZ, J.J., LUJAN, J.R., Programas de Recuperación de los Desechos Radiactivos de las Trincheras y de Descontaminación del Predio del Centro de Almacenamiento de Desechos Radiactivos, PORIT.DOC 16/07/99, Instituto Nacional de Investigaciones Nucleares, Mexico City (1999).
[186] NUCLEAR DECOMMISSIONING AUTHORITY, Radioactive Wastes in the UK: Context and Methodology Report, NDA, Moor Row (2017).
[187] NUCLEAR DECOMMISSIONING AUTHORITY, Draft Strategy: Published August 2020 for Consultation, West Row, Cumbria, Nucelar Decommissioning Authority, NDA, Moor Row (2020).
[188] ENVIRONMENT AGENCY, Decision Document: Low Level Waste Repository, EA, Bristol (2015).
[189] ENVIRONMENT AGENCY, Review of LLW Repository Ltd's 2011 Environmental Safety Case: Overview Report, EA, Bristol (2015).
[190] ENVIRONMENT AGENCY, NORTHERN IRELAND ENVIRONMENT AGENCY, SCOTTISH ENVIRONMENT PROTECTION AGENCY, Near-surface Disposal Facilities on Land for Solid Radioactive Wastes Guidance on Requirements for Authorisation, EA, Bristol (2009).
[191] LOW LEVEL WASTE REPOSITORY, 2019–20 Environmental Monitoring Summary Report, Rep. LLWR-RP/S&E/09334.05/0050, LLWR, Holmrook (2020).
[192] WAREING, A.S., EDEN, L., JONES, A., BALL, M., LLWR Lifetime Project, The Inventory of Past and Future Disposals at LLWR, Rep. 9124, Issue 4, Nexia Solutions, Warrington (2011).
[193] LOW LEVEL WASTE REPOSITORY, LLWR Environmental Safety Case: Selected Retrievals, Rep. LLWRP 675, LLWR, Holmrook (2009).
[194] MATOSHKO, A., BUGAI, D., DEWIERE, L., SKALSKYY, A., Sedimentological study of the Chernobyl NPP site to schematise radionuclide migration conditions. Environ. Geol. **46** 6 (2004) 820–830, https://doi.org/10.1007/s00254-004-1067-3
[195] IL'ICHEV, V.A., AKHUNOV, V.D., Study of the Radioactive Waste Burial Sites, Development of the Technologies and Measures for their Long Term Localization and Re-disposal — Realization of Radiation and Hydrogeological Monitoring at the Burial Sites and Adjoining Territories, Institute of Industrial Technologies Moscow, Chernobyl (1992) (in Russian).
[196] KASHPAROV, V., et al., Radionuclide migration in the experimental polygon of the Red Forest waste site in the Chernobyl zone, Part 1: Characterization of the waste trench, fuel particle transformation processes in soils, biogenic fluxes and effects on biota, Appl. Geochem. **27** 7 (2012) 1348–1358, |https://doi.org/10.1016/j.apgeochem.2011.11.004

[197] BUGAI, D., DEWIERE, L., Global Model for the ^{90}Sr Transport in the Aquifer at Chernobyl Pilot Site: Model Calibration and Sensitivity Analyses, Rep. IRSN/DEI/SARG 04-17, Institute for Radioprotection and Nuclear Safety, Fontenay-aux-Roses (2004).

[198] ARDOIS, C., SZENKNECT, S., Strontium-90 Interaction Process in the 'Soil–solution' System of the Chernobyl Pilot Site, Rep. IRSN/DEI/SARG-04–15, Institute for Radioprotection and Nuclear Safety, Fontenay-aux-Roses (2004).

[199] DEWIERE, L., BUGAI, D., GRENIER, C., KASHPAROV, V., AHAMDACH, N., ^{90}Sr migration to the geo-sphere from a waste burial in the Chernobyl exclusion zone, J. Environ. Radioact. **74** 1 (2004) 139–150, https://doi.org/10.1016/j.jenvrad.2004.01.019

[200] BUGAI, D., et al., Characterization of subsurface geometry and radioactivity distribution in the trench containing Chernobyl clean-up wastes, Environ. Geol. **47** 6 (2005) 869–881, https://doi.org/10.1007/s00254-004-1218-6

[201] THIRY, Y., et al., Impact of Scots pine (*Pinus sylvestris* L.) plantings on long term ^{137}Cs and ^{90}Sr recycling from a waste burial site in the Chernobyl Red Forest, J. Environ. Radioact. **100** 12 (2009) 1062–1068, https://doi.org/10.1016/j.jenvrad.2009.05.003

[202] KASHPAROV, V.A., et al., Biogenic migration of radionuclides from 'Red Forest' waste dumps, Bull. Ecol. Status Exclusion Zone **1** (2002) 32–8 (in Ukrainian).

[203] DZHEPO, S.P., et al., "Hydrogeological Effects of the Principal Radioactive-Waste Burial Sites Adjacent to the Chernobyl NPP (Review)", Proc. Int. Symp. Shelter Safety '94, Zeleny Mys, 1994, NEA-OECD, Paris (1994), 370–382, https://doi.org/10.13140/2.1.4138.3684

[204] BUGAI, D.A., WATERS, R.D., DZHEPO, S.P., SKALSKIJ, A.S., Risks from radionuclide migration to groundwater in the Chernobyl 30-km zone, Health Phys. **71** 1 (1996) 9–18, https://doi.org/10.1097/00004032-199607000-00002

[205] MOLITOR, N., DRACE, Z., STANGE-WIEHR, I., The International Projects to Develop and Support Short and Long Term Measures at ChNPP Unit 4 Site (2019), https://www.ispnpp.kiev.ua/wp-content/uploads/2019/2019_32/c101.PDF

[206] MOLITOR, N. (Ed.), "Recent safety assessment findings on management of wastes from Chornobyl accident", Int. Conf. Nuclear Decommissioning and Environment Recovery, 2017, Slavutych (2017).

[207] UNITED STATES ENVIRONMENTAL PROTECTION AGENCY, Comprehensive Five-Year Review Guidance, Rep. EPA 540-R-01-007 OSWER No. 9355.7-038-P, US EPA, Washington, DC (2001).

[208] UNITED STATES ENVIRONMENTAL PROTECTION AGENCY, Guidance for Conducting Remedial Investigations and Feasibility Studies Under CERCLA, Rep. PE89-184626 EPA/540/G-89/004, OSWER Directive 9355.3-0, US EPA, Washington, DC (1988).

[209] ENVIRONMENTAL PROTECTION AGENCY, Superfund Record of Decision: Maxey Flats Nuclear Disposal, KY, First Remedial Action – Final, Rep. EPA/ROD/R04-91/097, US EPA, Washington, DC (1991).

[210] US ENVIRONMENTAL PROTECTION AGENCY, (Draft) Technical Guidance for RCRA/CERCLA Final Covers, Rep. EPA 540-R-04-007 OSWER 9283.1-26, US EPA, Washington, DC (2004).

[211] NIKOLENKO, V.M., LEDENEV, A.I., KASHIRNY, A.N., Assessment of conditions of storage sites of radioactive wastes in Kiev Region, Problems of Chernobyl Exclusion Zone **2** (1995) 51.

[212] CLEVELAND, J.M., REES, T.F., Characterization of plutonium in Maxey Flats radioactive trench leachates, Science **212** 4502 (1981) 1506–1509, https://doi.org/10.1126/science.212.4502.1506

[213] KABIR, A., Safety Assessment of the Little Forest Legacy Site (LFLS), Rep. ANSTO/TN/N/2013-10 Rev. 1, ANSTO, Lucas Heights (2014).

[214] BUGAI, D., AVILA, R., Scenarios and ways of radioactive substances leakage from near-surface radioactive waste disposal points: a brief overview of historical cases, Nucl. Radiat. Saf. **3** 87 (2020) 21–27, https://doi.org/10.32918/nrs.2020.3(87).03

[215] THIRY, Y., et al., Impact of Scots pine (*Pinus sylvestris* L.) plantings on long term ^{137}Cs and ^{90}Sr recycling from a waste burial site in the Chernobyl Red Forest, J. Environ. Radioact. **100** 12 (2009) 1062–1068, https://doi.org/10.1016/j.jenvrad.2009.05.003

ABBREVIATIONS

AGE	Área de Gestión Ezeiza (Argentina)
ANSTO	Australian National Science and Technology Organisation
ARN	Nuclear Regulatory Authority (Argentina)
CADER	Centro de Almacenamiento de Desechos Radiactivos (Mexico)
CAE	Centro Atómico Ezeiza
ChNPP	Chornobyl nuclear power plant
CNEA	National Atomic Energy Commission of Argentina
CNSNS	Comisión Nacional de Seguridad Nuclear y Salvaguardias (Mexico)
CRL	Chalk River Laboratories
CSM	conceptual site model
EA	Environment Agency (United Kingdom)
ESC	Environmental Safety Case
HLW	high level radioactive waste
ILW	intermediate level radioactive waste
ICRP	International Commission on Radiological Protection
ISAM	Improvement of Safety Assessment Methodologies for Near Surface Disposal Facilities
LFLS	Little Forest Legacy Site (Australia)
LLW	low level radioactive waste
LLWR	Low Level Waste Repository (United Kingdom)
RWTSS	Radioactive Waste Temporary Storage Site (Ukraine)
US DOE	United States Department of Energy
US EPA	United States Environmental Protection Agency

CONTRIBUTORS TO DRAFTING AND REVIEW

Booth, P.	Consultant, United Kingdom
Bugai, D.	Institute of Geological Sciences, Ukraine
Clarke, R.	Environment Agency, United Kingdom
Cruickshank, J.	Sellafield Ltd, United Kingdom
Miller, S.	Canada
Monken-Fernandes, H.	International Atomic Energy Agency
Ormai, P.	Hungary
Payne, T.	Australian Nuclear Science and Technology Organisation, Australia
Punt, A.	Consultant, United Kingdom
Richards, J.	Environmental Protection Agency, United States of America
Shevelan, J.	Low Level Waste Repository Ltd, United Kingdom
Shorter, K.	Canadian Nuclear Laboratories, Canada
Suárez Alvarado, R.	Comisión Nacional de Seguridad Nuclear y Salvaguardias, Mexico
Tangir, D.	Comisión Nacional de Energia Atómica, Argentina
Yankovich, T.	International Atomic Energy Agency

Technical Meetings

Sellafield, United Kingdom: 12–16 September 2016
Kiev, Ukraine: 22–26 May 2017
Vienna, Austria: 23–26 October 2018
Sydney, Australia: 26–30 August 2019

Consultants Meetings

Vienna, Austria: 6–10 July 2015, 30 November–1 December 2017, 22–26 October 2018, 13–17 July 2020
Sellafield, United Kingdom: 12–16 September 2016
Sydney, Australia: 26–30 August 2019

Structure of the IAEA Nuclear Energy Series*

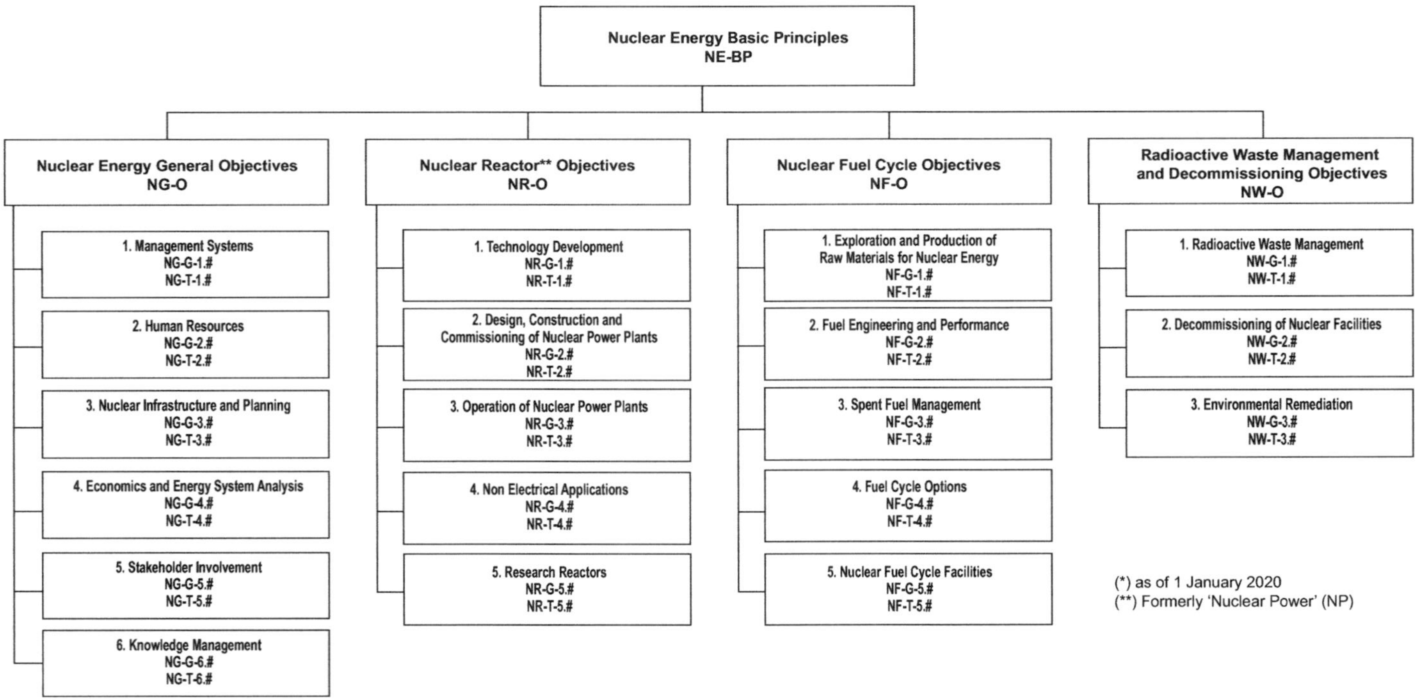

(*) as of 1 January 2020
(**) Formerly 'Nuclear Power' (NP)

Key

BP: Basic Principles
O: Objectives
G: Guides and Methodologies
T: Technical Reports
Nos 1–6: Topic designations
#: Guide or Report number

Examples

NG-G-3.1: Nuclear Energy General (**NG**), Guides and Methodologies (**G**), Nuclear Infrastructure and Planning (topic **3**), **#1**
NR-T-5.4: Nuclear Reactors (**NR**), Technical Report (**T**), Research Reactors (topic **5**), **#4**
NF-T-3.6: Nuclear Fuel (**NF**), Technical Report (**T**), Spent Fuel Management (topic **3**), **#6**
NW-G-1.1: Radioactive Waste Management and Decommissioning (**NW**), Guides and Methodologies (**G**), Radioactive Waste Management (topic **1**) **#1**

IAEA
International Atomic Energy Agency

CONTACT IAEA PUBLISHING

Feedback on IAEA publications may be given via the on-line form available at:
www.iaea.org/publications/feedback

This form may also be used to report safety issues or environmental queries concerning IAEA publications.

Alternatively, contact IAEA Publishing:

Publishing Section
International Atomic Energy Agency
Vienna International Centre, PO Box 100, 1400 Vienna, Austria
Telephone: +43 1 2600 22529 or 22530
Email: sales.publications@iaea.org
www.iaea.org/publications

Priced and unpriced IAEA publications may be ordered directly from the IAEA.

ORDERING LOCALLY

Priced IAEA publications may be purchased from regional distributors and from major local booksellers.

23-05494E-T

FSC
www.fsc.org
MIX
Papier | Fördert
gute Waldnutzung
FSC® C083411

Zeitfracht Medien GmbH
Ferdinand-Jühlke-Straße 7
99095 Erfurt, Deutschland
produktsicherheit@kolibri360.de